Numerical Approaches to Optimal Control: Tackling Nonlinear Systems and Constraints

Nama

Contents

1 Introduction

1

1.1 Literature review on stochastic model predictive control 1

    1.1.1 General formulation in SMPC 5

    1.1.2 Theoretical approaches in SMPC 11

    1.1.3 Distributed stochastic MPC 21

1.2 Literature review on event-based MPC 23

    1.2.1 Self-triggered SMPC 23

1.3 Motivation and organization of Ph.D. book 25

2 Stochastic Self-triggered MPC for Linear Constrained Systems under Additive Uncertainty and Chance Constraints **27**

2.1 Introduction 27

2.2 Problem formulation 30

    2.2.1 Reformulation of the chance constraints 33

    2.2.2 Cost function 35

    2.2.3 Terminal constraint 36

    2.2.4 Initial constraints 36

2.3 Stochastic self-triggered MPC problem 37

2.4 Feasibility and stability analysis 38

2.5 Numerical examples 45

    2.5.1 Comparison between the self-triggered and the periodically triggered stochastic MPC 45

    2.5.2 2-D point-mass double-integrator plant 48

2.6 Conclusions 50

3 Stochastic Self-triggered MPC with Adaptive Prediction Horizon for Linear Systems subject to Chance Constraints   **51**

3.1 Introduction 51

3.2 Problem setup 55

    3.2.1 System dynamics and chance constraints 55

    3.2.2 Self-triggered mechanism 55

3.3 Cost function and chance constraints handling 58

    3.3.1 Receding horizon performance with terminal cost function . . 58

    3.3.2 Chance constraints handling 62

    3.3.3 Terminal cost and terminal constraints 66

3.4 Stochastic self-triggered MPC with adaptive prediction horizon 68

    3.4.1 Optimization problem and algorithm 68

    3.4.2 Closed-loop properties 70

3.5 Numerical examples 72

3.6 Conclusions 76

4 Distributed Self-triggered Stochastic MPC for CPSs with Coupled Chance Constraints: A Stochastic Tube Approach   **77**

4.1 Introduction 77

4.2 Problem formulation 80

    4.2.1 Distributed cyber-physical control systems 80

    4.2.2 Self-triggered mechanism overview 82

4.3 Centralized Self-triggered SMPC problem under coupled chance constraints 85

    4.3.1 Cost function reformulation 86

    4.3.2 Local and coupled chance constraints handling 90

    4.3.3 Centralized self-triggered SMPC algorithm 95

4.4 Distributed self-triggered stochastic MPC under coupled chance constraints 98

    4.4.1 Distributed self-triggered SMPC algorithm 98

    4.4.2 Closed-loop properties of the distributed algorithm 100

4.5 Numerical examples 104

    4.5.1 Case 1: Homogeneous subsystems 104

    4.5.2 Case 2: Heterogeneous subsystems 107

4.6 Conclusions 108

5 Conclusions   **109**

5.1 Conclusions 109

Chapter 1

Introduction

In this chapter, the merits and challenges of stochastic model predictive control (SMPC) are first introduced. Then existing methodologies and research results on distributed SMPC and self-triggered SMPC are reviewed. Finally, the motivation of conducting the research of this **book** and the overall **book** outline is presented.

1.1 Literature review on stochastic model predictive control

Optimal control has been widely applied to modern control systems design and has drawn great attention for decades. In the optimal control theory, the control problem is formulated as an optimization problem, and the control law is calculated by solving the optimal control problem (OCP). Comparing with traditional control methods like PID control, optimal control is capable of providing an optimal control law in a systematic way. However, optimal control can achieve an analytic expression of the optimal control law only for some relatively simpler cases, for instance, unconstrained linear systems. Specifically, the optimal feedback control law for unconstrained linear systems with a quadratic cost function is in a simple linear form, and the optimal control gain is obtained by solving a Riccati equation. In practical systems, most of the physical plants are essentially nonlinear systems subject to physical constraints. For these cases, it is difficult to obtain an analytical solution to the optimal control problems. In order to find the optimum to an intractable optimization problem, approximate solutions have to be taken into account. Due to the rapid development in computer technologies and advanced optimization algorithms, numerical solutions to nonlinear constrained optimal control problems have been extensively studied. Accordingly, model predictive control (MPC) is proposed to seek a sub-optimal solution to the practical optimal control problem.

MPC has attracted considerable research attention in the past decade due to its distinct advantages and broad applications in the industry. Numerous notable results have been proposed in the literature, and the research interest is still increasing in recent years. Applications of MPC in the industry have been reported in [1] and theoretical properties of MPC are well discussed in [2]. The philosophy of MPC in discrete paradigm is briefly illustrated in Figure 1.1. In Figure 1.1, the dash-dot

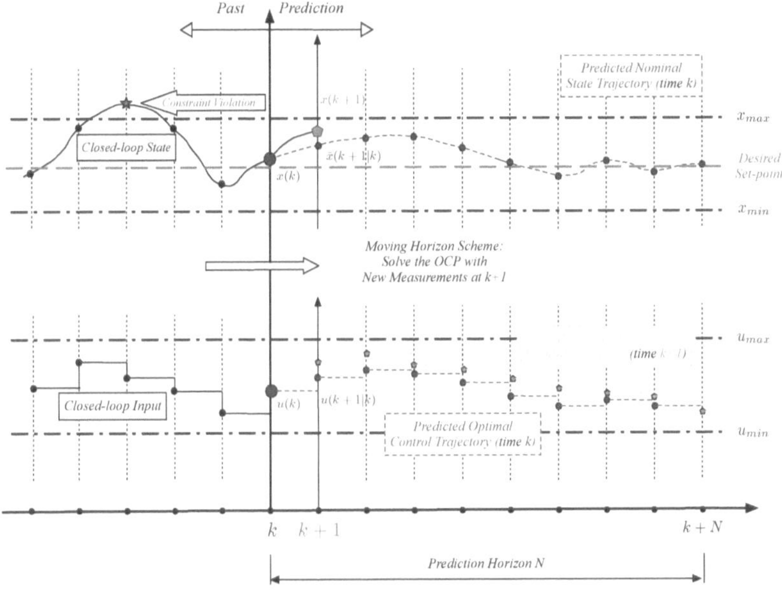

Figure 1.1: The schematic of model predictive control.

lines $x_{\max}$, $x_{\min}$, $u_{\max}$ and $u_{\min}$ represent the state constraints and input constraints, respectively. The black curves on the left hand side of the vertical axis are the real closed-loop state trajectory and the control input trajectory. The dash line on the top is the predicted nominal state trajectory, and the bottom one is the predicted optimal control trajectory. At time instant k, given the current system state $x(k)$ and a hypothetical control input sequence $\mathbf{u}(k) = \left[u(k|k), u(k+1|k), \ldots, u(k+N|k)\right]$, MPC utilizes the system dynamics to predict the system behavior N steps ahead. A constrained optimal control problem is then formulated where the control input sequence $\mathbf{u}(k)$ is defined as decision variables. The number of decision variables is related to the prediction horizon N, and the cost function can be defined as a sum of weighted matrix norms of predicted state and control input in general. Then, an optimal control input sequence $\mathbf{u}^*(k)$ can be obtained by solving the MPC problem online, and only the first element in $\mathbf{u}^*(k)$ is applied to the plant. The rest elements in $\mathbf{u}^*(k)$ are discarded and the process is repeated at the subsequent time $k+1$ with the new state measurement $x(k+1)$. Since the prediction horizon in MPC is shifted forward at each time instant, MPC is also called receding horizon control. One feature of MPC is that a compromise between the optimality and the computational load can be guaranteed by applying the receding horizon scheme.

It should be noted that feedback is implicitly introduced into MPC by the receding horizon scheme, and such a receding horizon fashion guarantees some certain degrees of inherent system robustness, as discussed in [3]. Rather than the inherent robustness of the MPC paradigm, elaborated investigation on the robustness of MPC under uncertainties and disturbances has drawn great attention. In practical systems, uncertainties are inevitable and arise in different ways. The system performance is greatly affected by the uncertainties. In the MPC area, robust model predictive control (RMPC) and SMPC have been respectively investigated to deal with model uncertainties and external disturbances in the system. In RMPC, each element in the set of uncertain parameters is treated without distinction. The critical feature of RMPC is to assume that the constraints are satisfied for all realizations of the uncertainty. Consequently, the worst-case consideration of the uncertainties is dominant in the RMPC controller design. However, from the practical application standpoint, some values of uncertainties are more likely to be than others. Intuitively, the chances of realizations of uncertainty lying around the nominal state are higher than that in the boundary of the uncertainty set. Alternatively, the model uncertainty can be viewed as stochastic with a known distribution. To utilize this statistic information, SMPC treats the model uncertainty from a statistical point of view.

When the system performance is strictly required, the constraints have to be imposed robustly to ensure that the constraints are satisfied with all realizations of uncertainties. In this case, we call the constraints as *hard constraints*. However, when constraint violations are permitted, it is conservative to impose the constraints robustly. With the price of allowing for constraint violations, the system performance can be improved, and the region of attraction can be enlarged. In this case, we call the constraints as *chance constraints* or *probabilistic constraints* since the constraints are permitted to be violated within a given probability. One unique feature in SMPC is to allow for constraint violations to utilize the probability distribution of uncertainties. An example of constraint violation is illustrated by the top-left red pentagram in the real closed-loop state trajectory in Figure 1.1. Different forms of chance constraints have been proposed in the literature, and details will be discussed in the sequel.

Both theoretical studies and practical applications of SMPC have been studied extensively in the literature over the past decade. An overview of SMPC applications in different areas is provided in Table 1.1. SMPC applications can be found in various emerging areas, such as the automotive industry, building climate control, microgrids, process industry and robotics. We consider building climate control as an example since it is the most widely studied SMPC applications in the literature. Buildings now consume about 40% of the total energy in the world and about 76% and 73% of electricity in Europe and the United States [4], respectively. In [4], SMPC is applied to the building climate control for the first time. A bilinear model is proposed to predict the state of the system, and decoupled time-varying chance constraints are considered in the problem formulation. The additive stochastic uncertainty comes from the estimation error in the weather prediction. In [5], the authors extend the ideas in [4] by using an affine disturbance feedback control law. In [6, 7], the nonlinear SMPC has been applied to HVAC systems. The SMPC is formulated as a nonlinear programming problem, which can be solved efficiently by the sequential quadratic programming method. In [6], based on Boole's inequality, the joint chance constraints in linear SMPC are decoupled, and the optimization problem can be solved by a tailored interior-point method. Based on the ideas of randomized optimization, a scenario-based SMPC approach has been used in the building climate control [8, 9] as well. Recently in [10, 11] new problem formulations have been proposed to reduce the amount of the sampled constraints in the scenario-based MPC.

Notations: $\mathbb{N}$ denotes the set of integers, and $\mathbb{N}_{[a,b]}$ represents the set of integers from a to b, where $a \leq b$, $a, b \in \mathbb{N}$. $\mathbb{N}_{>a}$, $\mathbb{N}_{\geq a}$, $\mathbb{N}_{<a}$, and $\mathbb{N}_{\leq a}$ denote sets $\{n \in \mathbb{N} | n > a\}$, $\{n \in \mathbb{N} | n \geq a\}$, $\{n \in \mathbb{N} | n < a\}$, $\{n \in \mathbb{N} | n \leq a\}$ for $a \in \mathbb{N}$. $\mathbb{R}^n$ stands for the n-dimensional real space.

Table 1.1: Applications of SMPC for linear and nonlinear systems.

	Stochastic tube	Scenario-based	Stochastic programming	Nonlinear SMPC
Automotive industry	[12, 13]	[14]	[15, 16, 17, 18]	
Building climate	[5, 6, 8]	[9, 10, 11, 19]		[6, 7]
Finance and operation policy	[20, 21]	[22, 23]		
Microgrids	[24, 25]	[26]		
Process industry	[27, 28, 29, 30, 31, 32]			[33, 34, 35, 36]
Robotics	[37, 38]	[39, 40]	[41]	

1.1.1 General formulation in SMPC

In this section, the most widely adopted problem formulations in SMPC algorithms design will be introduced. First, an appropriate model should be selected to describe the statistical information in the system. Then, chance constraints imposed on the states and control inputs can be suitably reformulated to a deterministic form. Finally, an optimal control problem with a suitable cost function and reformulated constraints can be solved to calculate the optimal control sequence.

The general system model considered in SMPC can be described by the following discrete-time model

$$x(k + 1) = f(x(k), u(k), w(k)), \tag{1.1}$$

where $x(k) \in \mathbb{R}^{n_x}, u(k) \in \mathbb{R}^{n_u}$ are the system state and control input, respectively. The function $f(\cdot)$ is a nonlinear Borel-measurable function that characterizes the system dynamics. $w(k) \in \mathbb{R}^{n_w}$ denotes the stochastic disturbance or uncertainty whose realization is unknown at current and future time instants.

In SMPC, it is always assumed that the probability distributions $\mathbf{P}_w$ of $w(k)$ is known, and assumptions on $w(k)$ play a vital role in SMPC controller design. In most existing literature in SMPC, $w(k)$ is assumed to lie in a bounded and convex nonempty set $\mathcal{W}$. In some papers, $w(k)$ is assumed to have some specific probability distributions or only the moments of the probability distribution are required. Few methods allow for general probability distributions due to the complexity in handling chance constraints and guaranteeing closed-loop properties. One common assumption

Table 1.2: SMPC classification: Representative model dynamics and uncertainty types.

	Linear system			Nonlinear system	
	Add.	Multi.	Add./Multi.	Time invariant	Time varying
Bounded	[42, 43, 44]	[45, 46, 47]	[48, 49]	[50, 34]	[51, 52]
Unbounded	[53, 54, 55, 56, 57, 30]	[58]			[59, 41]

in SMPC is that the disturbance elements $w(k+i), i \in \mathbb{N}_{\geq 0}$ are independent and identically distributed. In Table 1.2, we categorize SMPC algorithms in terms of different assumptions on system dynamics and uncertainty types. Much of the SMPC algorithms in the literature have been developed for linear systems. For linear systems with additive disturbances, the general formulation in (1.1) is modified as

$$x(k+1) = Ax(k) + B_u u(k) + B_w w(k), \tag{1.2}$$

where A, B_u and B_w are state equation matrices with appropriate dimension. For linear systems with multiplicative uncertainties, the model is described as

$$x(k+1) = Ax(k) + B_u u(k) + \sum_{j=1}^{q}[A_j x(k) + B_j u(k)]w_j(k), \tag{1.3}$$

where time-varying uncertainties in state matrices A, B_u are represented by the sequence $\{w_1(k), \ldots, w_j(k), \ldots, w_q(k)\}$.

The general form of feedback control policy in SMPC can be defined as

$$\boldsymbol{\pi}(\cdot) := \{\pi_0(\cdot), \ldots, \pi_{N-1}(\cdot)\}, \tag{1.4}$$

where the function $\pi_i(\cdot) : \mathbb{R}^{(i+1)n_x} \to \mathbb{U}, i \in \mathbb{N}_{[0,N-1]}$, denotes a general feedback control law and $N \in \mathbb{N}_{>0}$ denotes the prediction horizon. Thus, the i-step ahead predicted control input $u(i|k)$ can be selected as $u(i|k) = \pi_i(\cdot)$. Different parameterization methods of (1.4) have been studied in SMPC, and we will discuss this topic in the next section. In general, the cost function in SMPC can be defined as

$$J_N(x(k), \pi) := \mathbb{E}_k \left\{ \sum_{i=0}^{N-1} l(x(i|k), u(i|k)) + l_f(x(N|k)) \right\}, \tag{1.5}$$

where the functions $l(\cdot, \cdot) : \mathbb{R}^{n_x} \times \mathbb{U} \to \mathbb{R}$ and $l_f(\cdot) : \mathbb{R}^{n_x} \to \mathbb{R}$ are defined as stage cost and terminal cost. $\mathbb{E}_k\{\cdot\}$ denotes the expectation of a random variable at time instant k. The i-step ahead predicted state and control input are defined as $x(i|k)$ and $u(i|k)$ given the initial state $x(k)$, feedback control law π, and probability distribution of $\{w(k), \ldots, w(k+i-1)\}$ at time instant k. In the SMPC framework, the predicted state $x(i|k), i \in \mathbb{N}_{[1,N]}$ are random variables affected by the uncertainties $w(k+i)$ in the system, implying that the cost function $J_N(x(k), \pi)$ contains stochastic components. Thus, the cost function in SMPC needs to be optimized in a probabilistic form and four different types of cost functions have been developed in the literature:

(J_1) Based on the certainty equivalence principle [60, 55, 52], the cost function in (1.5) can be rewritten as a deterministic one:

$$J_N(x(k), \pi) := \sum_{i=0}^{N-1} l(\bar{x}(k+i|k), u(k+i)) + l_f(\bar{x}(k+N)),$$

where the nominal state $\bar{x}(i|k)$ evolves according to system dynamics $\bar{x}(i+1|k) = f(\bar{x}(i|k), u(i|k), \bar{w}(k+i))$ with initial condition $\bar{x}(k|k) = x(k)$. The term $\bar{w}(k+i)$ is the nominal disturbance trajectory which is usually defined as the expected value of $w(k+i)$.

(J_2) For linear systems, one commonly adopted cost function used in the literature [58, 56] is defined in linear quadratic form as follows:

$$J_N(x(k), \pi) := \mathbb{E}_k \left\{ \sum_{i=0}^{N-1} (\|x(i|k)\|_Q^2 + \|u(i|k)\|_R^2) + \|x(N|k)\|_P^2 \right\},$$

where the cost function (1.5) is reformulated as a function of mean and variance of the state and control input variables. Let $\text{Var}\{\cdot\}$ denote the variance of a random variable. If we define $\mathbb{E}[x(i|k)] = \bar{x}(i|k)$, $\mathbb{E}[u(i|k)] = \bar{u}(i|k)$, $\text{Var}(X(i|k)) = X(i|k)$ and $\text{Var}(u(i|k)) = U(i|k)$, then the cost can be expressed as

$$J_N(x(k), \pi) := \sum_{i=0}^{N-1} \|\bar{x}(i|k)\|_Q^2 + \|\bar{u}(i|k)\|_R^2 + \|\bar{x}(N|k)\|_P^2$$

$$+ \sum_{i=0}^{N-1} \mathrm{tr}(QX(i|k) + RU(i|k)) + \mathrm{tr}(PX(k+N)).$$

(J$_3$) Another type of cost function for linear systems is the expectation infinite-horizon cost function [43, 45] defined as follows:

$$J_\infty(x(k), \pi) := \mathbb{E}_k\left\{ \sum_{i=0}^{\infty} \|x(i|k)\|_Q^2 + \|u(i|k)\|_R^2 \right\}.$$

(J$_4$) In the scenario-based method [61, 62], a sampled average cost function over N_s uncertainty realizations can be formulated as follows:

$$J \simeq \frac{1}{N_s}\sum_{l=1}^{N_s} \sum_{i=0}^{N-1} l(x^{[l]}(i|k), u(i|k)) + l_f(x^{[l]}(k+N)).$$

Define $w^{[l]}(k+i), i \in \mathbb{N}_{[0,N-1]}, l \in \mathbb{N}_{[1,N_s]}$ as the lth disturbance realization at predicted time instant $k+i$ and N_s is the number of samples. The i-step ahead predicted state $x^{[l]}(i|k)$ evolves according to $x^{[l]}(i+1|k) = f(x^{[l]}(i|k), u(i|k), w^{[l]}(k+i))$ with initial condition $x^{[l]}(0|k) = x(k)$.

The optimization problem in SMPC is commonly performed subject to chance constraints. The general form of joint chance constraints [31] on the predicted state $x(i|k)$ over the prediction horizon is given by:

$$\Pr_k\left[g_j(x(i|k)) \le 0, j \in \mathbb{N}_{[1,N_x]} \right] \ge 1 - p, i \in \mathbb{N}_{[0,N-1]}, \tag{1.6}$$

where $g_j(\cdot) : \mathbb{R}^{n_x} \to \mathbb{R}$ is a Borel-measurable function, N_x is the number of constraints and p is the probability of constraint violations. The value of p provides a tradeoff between the system performance and chance of constraints violation. For all constraints $j \in \mathbb{N}_{[1,N_x]}$ over the prediction horizon $i \in \mathbb{N}_{[0,N-1]}$, the probability $\Pr_k$ is conditional on the system state at time k. One reason to impose the chance constraints on the state is that, when the disturbance $w(k)$ in (1.1) is unbounded, it may result in unavoidable constraint violations since the disturbance could be arbitrarily large. Moreover, although the disturbance can be assumed to be bounded, the worst-case consideration in RMPC may lead to a conservative result and the system performance can be improved by taking the stochastic form of (1.6) into account. Due to the difficulty of expressing the joint chance constraints in a tractable way, one simple method [63] is to approximate the overall set using a sequence of individual chance constraints. An overview of this problem can be found in [64]. As demonstrated in [57], this approach is relatively conservative, and the resulting feasible region is much smaller than the one obtained by using joint chance constraint directly. Thus, in most of existing literature on SMPC, only individual chance constraints in the following form are considered:

$$\mathrm{Pr}_k\Big[g_j(x(i|k)) \leq 0\Big] \geq 1 - p_j, j \in \mathbb{N}_{[1,N_x]}, i \in \mathbb{N}_{[0,N-1]}, \tag{1.7}$$

where p_j is the probability of constraint violation for different inequality constraints. In the literature, three types of formulation of (1.7) have been proposed:

(C_1) Individual chance constraints defined in (1.7) represent the probability of constraints violation for pointwise-in-time constraints. It is the most commonly adopted type of chance constraints.

(C_2) Average constraints violation formulation in [48, 46, 42] is defined as the number of constraint violations over a horizon N_h will not exceed a given maximum number $N_{\max}$:

$$\frac{1}{N_h} \sum_{t=0}^{N_h-1} \mathbf{1}_{g_j}(x(i|k)) < \frac{N_{\max}}{N_h}, j \in \mathbb{N}_{[1,N_x]}, i \in \mathbb{N}_{[0,N-1]}.$$

The indicator function $\mathbf{1}_{g_j}(x(i|k))$ is defined as

$$\mathbf{1}_{g_j}(x(i|k)) := \begin{cases} 1, & g_j(x(i|k)) > 0; \\ 0, & g_j(x(i|k)) \leq 0. \end{cases}$$

(C_3) Expectation type [58, 65] of chance constraints is defined as

$$\mathbb{E}[g_j(x(i|k))] \leq 0, j \in \mathbb{N}_{[1,N_x]}, i \in \mathbb{N}_{[0,N-1]}.$$

In this formulation, it is required that the constraints are satisfied on average, and hence constraint violations are not considered explicitly.

The effect and comparison of these types of stochastic formulation of chance constraints in SMPC can be found in [53]. An integrated-type chance constraint in [66] provides a more quantitative way to express the idea of constraint violation but receives relatively little attention. A summary of representative works using these types of constraint formulation and cost function is given in Table 1.3.

Table 1.3: Representative types of chance constraints and cost function in SMPC.

	Formulation	Linear	Nonlinear
State constraints			
Joint-type	$\Pr_k\left[g_j(x(i\|k)) \leq 0, j \in \mathbb{N}_{[1,N_x]}\right] \geq 1 - p$	[31]	
Individual-type	$\Pr_k\left[g_j(x(i\|k)) \leq 0\right] \geq 1 - p_j$	[43], [45]	[50]
Average violation type	$\frac{1}{N_\mathrm{h}}\sum_{t=0}^{N_\mathrm{h}-1} \mathbf{1}_{g_j}(x(i\|k)) < \frac{N_{\max}}{N_\mathrm{h}}$	[48, 46, 42]	
Expectation-type	$\mathbb{E}[g_j(x(i\|k))] \leq 0$	[58]	[65]
Input constraints			
Hard input type	$h(x(i\|k), u(i\|k)) \leq 0$	[34]	[35]
Probabilistic input type	$\Pr_k\{h(x(i\|k), u(i\|k)) \leq 0\} \geq 1 - p_u$	[58],[67]	
Saturated input type	$\|h(u(i\|k))\|_\infty \leq u_{max}$	[68]	
Cost function			
Equivalence type	$\sum_{i=0}^{N-1} l(\bar{x}(i\|k), u(i\|k)) + l_f(\bar{x}(N\|k))$	[55, 60]	
with terminal penalty	$\mathbb{E}_k\left\{ \sum_{i=0}^{N-1} \|x(i\|k)\|_Q^2 + \|u(i\|k)\|_R^2 + \|x(N\|k)\|_P^2 \right\}$	[58, 56]	
Infinite horizon type	$\mathbb{E}_k\left\{ \sum_{i=0}^{\infty} \|x(i\|k)\|_Q^2 + \|u(i\|k)\|_R^2 \right\}$	[45, 43]	
Sampled average type	$\frac{1}{N_s}\sum_{j=1}^{N_s} \sum_{i=0}^{N-1} l(x^{[l]}(i\|k), u(i\|k)) + l_f(x^{[l]}(N\|k))$	[52]	

The stochastic optimal control problem for system (1.1) subject to both probabilistic state constraints (1.7) and hard input constraints can be formulated as

$$
\begin{aligned}
J_N^*(x(k)) :=\ & \min_{\pi} J_N(x(k), \pi) \\
\text{s.t.}\quad & \bar{x}(0|k) = x(k), \\
& \bar{x}(i+1|k) = f(\bar{x}(i|k), u(i|k), w(k+i)), \quad && i \in \mathbb{N}_{[0,N-1]} \\
& \pi_i(\cdot) \in \mathbb{U}, && i \in \mathbb{N}_{[0,N-1]} \\
& \mathrm{Pr}_k\Big[g_j(x(i|k)) \le 0\Big] \ge 1 - p_j, && j \in \mathbb{N}_{[1,N_x]}, i \in \mathbb{N}_{[0,N-1]}, \\
& w(k+i) \sim \mathbf{P}_w, && i \in \mathbb{N}_{[0,N-1]},
\end{aligned}
$$

$$(1.8)$$

where $J_N^*(x(k))$ is the optimal cost function given the optimal control law $\boldsymbol{\pi}(\cdot) = \Big[\pi_0(\cdot) \ldots \pi_{N-1}(\cdot)\Big]$. By solving the stochastic OCP (1.8) at each time instant k, the optimal control action $u(k) = \pi_0^*(x(k|k))$ will be applied to the plant, which implies the receding horizon implementation of SMPC.

1.1.2   Theoretical approaches in SMPC

As discussed in Section 1.1.1, the general SMPC problem can be casted as an chance-constrained stochastic OCP as shown in (1.8). In order to solve the stochastic OCP (1.8), there are three main challenges: a) The form of control law π is arbitrary; b) the implementation of chance constraints makes the OCP (1.8) intractable in general; c) the propagation of uncertainties through system dynamics is complex, especially when considering nonlinear systems. Numerous theoretical approaches have been proposed in the literature to generate a tractable surrogate for OCP (1.8). Specifically, four main approaches have been proposed to approximate the chance-constrained stochastic OCP (1.8). In Table 1.4, we categorize SMPC algorithms in terms of different control parameterization methods and uncertainty propagation methods.

(A_1) *Stochastic tube or analytic approximation approach for linear systems.*

In order to obtain predicted system states $x(i|k)$ in the prediction horizon, we need to determine the probability distribution of system states $x(i|k)$ over multiple time instants $i \in \mathbb{N}_{[0,N-1]}$. This requires evaluating a multivariable convolution integral, and in general, this problem is intractable for systems of large dimensions, especially for nonlinear systems or multiplicative uncertainty. The chance-constrained stochastic OCP (1.8) is reformulated as deterministic terms thanks to the superposition property in linear systems. For linear systems, dynamics can be decomposed into nominal dynamics and error dynamics,

Table 1.4: Control parameterization and uncertainty propagation

| | Uncertainty propagation | | | | | |
| | Stochastic tube | | Scenario-based | NSMPC | | |
	Mixed	Polytope		gPCEs	FP	GM
State-feedback	[43]	[45, 46, 48, 69]	[49]	[50]	[35]	[70, 71]
Disturbance-feedback	[55]	[42]				
Output-feedback	[44]	[67]				

both of which can be tackled separately. This reformulation will result in some conservativeness, and the offline design is often cumbersome.

In stochastic tube approaches [48, 46, 45, 69, 43, 72], the stochastic OCP is defined as the infinite horizon cost function $J_\infty(x(k), \pi)$ subject to chance constraints. The construction of stochastic tubes can ensure closed-loop properties such as recursive feasibility of the optimization algorithm, chance constraint satisfaction, and system stability. Generally speaking, the control law (1.4) is parameterized by the *dual mode prediction paradigm*, which consists of the state feedback control policy $u(i|k) = Kx(i|k) + c(i)$ with perturbation variables $c(i)$ for $i \in \mathbb{N}_{[0,N-1]}$ and a pre-stabilizing control law $u(i|k) = Kx(i|k)$ for $i \in \mathbb{N}_{\geq N}$. The sequence of control perturbation variables $\{c(0), \ldots, c(N-1)\}$, being decision variables, is calculated by solving the stochastic OCP (1.8). The feedback gain matrix K is designed by ensuring that $\Phi := A + B_u K$ to be Schur stable. The importance of introducing the pre-stabilizing control law $u(i|k) = Kx(i|k)$ is in three-folds: a) The mean-square stability of the system $x(i+1|k) = \Phi x(i|k) + B_w w(k+i)$ without constraints can be guaranteed under the control law $u(i|k) = Kx(i|k)$; b) based on the convergence analysis of $x(i|k)$ under the control law $u(i|k) = Kx(i|k)$ for $i \in \mathbb{N}_{\geq N}$, the infinite-horizon cost $J_\infty(x(k), \pi)$ can be reformulated into a finite horizon cost as $J_\infty(x(k), \pi) - L_{ss}$, where $L_{ss} = \lim_{i \to \infty} \mathbb{E}_k \{ \|x(i|k)\|_Q^2 + \|u(i|k)\|_R^2 \}$ is the limit of stage cost function; c) the terminal invariant set is constructed based on the control law $u(i|k) = Kx(i|k)$ to guarantee the recursive feasibility of the algorithm.

To demonstrate constraints tightening in stochastic tube approaches, we consider linear systems with bounded additive uncertainties in the form of (1.2) as an example, where disturbance $w(k)$ is assumed to be a white noise with zero mean. Due to the superposition principle, we can decompose the real system state $x(i|k)$ into nominal state $\bar{x}(i|k)$ and error state $e(i|k)$:

$$
\begin{aligned}
x(i|k) &= \bar{x}(i|k) + e(i|k), i \in \mathbb{N}_{\geq 0}, \\
\bar{x}(i+1|k) &= \Phi\bar{x}(i|k) + B_u c(i), \\
e(i+1|k) &= \Phi e(i|k) + B_w w(k+i),
\end{aligned}
\tag{1.9}
$$

with $\bar{x}(0|k) = x(k)$ and $e(0|k) = 0$. $e(i|k)$ describes the effect of uncertainties on the predicted state $x(i|k)$. Due to this decomposition, the state linear chance constraints $\mathrm{Pr}_k\{g^\mathrm{T}x(i|k) \leq h\} \geq 1 - p$ can be replaced by hard tightened constraint $g^\mathrm{T}\bar{x}(i|k) \leq \hat{h}, \hat{h} \leq h$ [43], such that $\mathrm{Pr}_k\{g^\mathrm{T}x(i|k) \leq h\} \geq 1-p$ is ensured for all $i \in \mathbb{N}_{[0,N-1]}$. The original chance constraints are approximated by tightened constraints on predicted nominal state $\bar{x}(i|k)$, leading to a reduction in the number of decision variables. The stochastic OCP is reformulated as a convex quadratic programming problem such that the computational complexity of stochastic tube SMPC problem is similar to that of a nominal MPC. Numerous variations of stochastic tube approaches, such as ellipsoidal tube [45], nested layer tube [46], tube with a fixed cross-section and varying scalings [69], tube with striped structure control policy [42], for either additive disturbances or multiplicative uncertainties have been studied in this area. It should be noted that the tube cross-section and scalings are computed offline. The efficiency of the SMPC algorithm relies on the construction of the tube section, which is essentially determined by the propagation of uncertain components through system dynamics.

(A$_2$) *Affine parameterization of control policy for linear systems.*

One of the key challenges in formulating a tractable stochastic OCP (1.8) is that the optimization over the arbitrary control policy $\pi(\cdot)$ is generally intractable. In general, the resulting stochastic OCP is a nonconvex problem, and various SMPC algorithms have been proposed to find a tractable surrogate of this problem. Different from the dual-mode prediction diagram used in stochastic tube approaches, affine state parameterization and affine disturbance parameterization have also been used in SMPC algorithm design to obtain a convex OCP. In contrast to stochastic tube approaches, the uncertainty $w(k)$ is generally assumed to be unbounded in this type of SMPC algorithm, while hard input constraints are imposed. This poses theoretical challenges on establishing the recursive feasibility of the SMPC algorithm and the stability of the system. To deal with these problems, the control policy can be affinely parameterized by a saturation function. In the following, we will illustrate these affine feedback control policies in the presence of both hard input constraint and unbounded uncertainties. For simplicity, the error dynamics for linear systems subject to additive disturbances evolves in a general form of

$$e(i|k) = E_i \mathbf{w}(k+i), \tag{1.10}$$

where $\mathbf{w}^{\mathrm{T}}(k+i) = [w(k)^{\mathrm{T}} \cdots w(k+i-1)^{\mathrm{T}}]$ and the vector E_i depends on the parameterization of control policy.

Open-loop policy. In [63, 54], an open-loop control policy $u(i|k) = c(i), i \in \mathbb{N}_{[0,N-1]}$ is utilized and the resulting optimization problem doesn't depend on the uncertainty sequence $\mathbf{w}(k+i)$. The vector E_i in (1.10) corresponds to the ith row of the matrix

$$\begin{bmatrix} B_w & 0 & \dots & 0 \\ AB_w & B_w & \dots & 0 \\ \vdots & \vdots & \ddots & \vdots \\ A^{N-1}B_w & A^{N-2}B_w & \dots & B_w \end{bmatrix}.$$

Under this type of control policy, the evolution of $e(k+i)$ is substantially uncontrolled. Even though this approach can be readily designed and implemented, the drawbacks of this approach are obvious: When the system is unstable, the strategy may cause the critical infeasible solution and even leads to the unstability of the closed-loop system.

State or error feedback policy. The control policy is called error feedback policy if it is defined in form of $u(i|k) = K_{i|k}e(i|k) + c(i), i \in \mathbb{N}_{[0,N-1]}$. Compared to state feedback control policy, the sequence of decision variables $\{c(0), K_{0|k}, \dots, c(N-1), K_{N-1|k}\}$ contains a larger set of decision variables. Under this control policy, the vector E_i is defined as the ith block row of the matrix

$$\begin{bmatrix} B_w & 0 & \cdots & 0 \\ \Phi_{k+1}B_w & B_w & \cdots & 0 \\ \vdots & \vdots & \ddots & \vdots \\ \Phi_{k+N-1}\cdots\Phi_{k+1}B_w & \Phi_{k+N-1}\cdots\Phi_{k+2}B_w & \cdots & B_w \end{bmatrix}. \qquad (1.11)$$

Alternatively, the control policy is called state feedback policy if it is defined as $u(i|k) = K_{i|k}x(i|k) + c(i), i \in \mathbb{N}_{[0,N-1]}$, as suggested in [67]. The hard input constraints are relaxed to probabilistic input constraints in [67] for linear systems subject to unbounded disturbances. Based on Cantelli's inequality, chance constraints are reformulated into deterministic forms. The sequence $\{c(0), K_{0|k}, \ldots, c(N-1), K_{N-1|k}\}$ is optimized by solving a convex OCP online to minimize the variance of state $x(i|k)$. Initial and terminal constraints are imposed to guarantee the feasibility of the resulting SMPC problem in the presence of the unbounded uncertainties, and the closed-loop system is shown to be input-state stable. The inherent limitation of this approach comes from the utilization of Cantelli's inequality that gives rise to a conservative approximation for chance constraint. Meanwhile, the optimization problem is nonconvex due to the control parameterization. If we further assume $K_{i|k} = K$, then E_i corresponds to the ith block-row of the matrix (1.11) with $\Phi_{k+i} = \Phi$. This type of control policy is widely utilized in stochastic tube approaches and the number of decision variables in solving the stochastic OCP (1.8) is finite.

Disturbance feedback control. In this approach [42, 55], the disturbance feedback control policy is defined as $u(i|k) = c(i) + \Theta_k\mathbf{w}(k+i), i \in \mathbb{N}_{[0,N-1]}$, where Θ_k is the ith row of matrix

$$\begin{bmatrix} 0 & 0 & 0 & \cdots & 0 \\ \theta_{k+1,k} & 0 & 0 & \cdots & 0 \\ \theta_{k+2,k} & \theta_{k+2,k+1} & 0 & \cdots & 0 \\ \vdots & \vdots & \vdots & \ddots & \vdots \\ \theta_{k+N-1,k} & \theta_{k+N-1,k+1} & \theta_{k+N-1,k+2} & \cdots & 0 \end{bmatrix}. \qquad (1.12)$$

The decision variables to (1.8) are the sequence of $\{c(0), \ldots, c(N-1)\}$ and elements in matrix (1.12). The error dynamics is described by

$$e(i+1|k) = Ae(i|k) + B_u c(i)\Theta_k \mathbf{w}(k+N-1) + B_w w(k+i), i \in \mathbb{N}_{[0,N-1]}.$$

Hence, the matrix E_i corresponds to the ith block-row of the matrix

$$\begin{bmatrix} B_w & 0 & \cdots & 0 \\ AB_w & B_w & \cdots & 0 \\ \vdots & \vdots & \ddots & \vdots \\ A^{N-1}B_w & A^{N-2}B_w & \cdots & B_w \end{bmatrix} + \begin{bmatrix} B_u & 0 & \cdots & 0 \\ AB_u & B_u & \cdots & 0 \\ \vdots & \vdots & \ddots & \vdots \\ A^{N-1}B_u & A^{N-2}B_u & \cdots & B_u \end{bmatrix} \Theta_k.$$

It shows that the uncertainty sequence $\mathbf{w}(k+i)$ is controlled by Θ_k, which can be obtained by solving the optimization problem online. The essential idea of the disturbance feedback control policy originates from the assumption: The realization of disturbance and predicted state will be known at a future time instant. Accordingly, such known information will be utilized to design predicted control input over the prediction horizon N. When unbounded uncertainties and hard input constraints are taken into account, the disturbance feedback control policy is further extended to a more general form of $u(i|k) = c(i) + \Theta_k \Psi_i(\mathbf{w}(k+i)), i \in \mathbb{N}_{[0,N-1]}$, as shown in [73], where $\Psi_i(\cdot) : \mathbb{R}^{i*n_w} \to \mathbb{R}^{i*n_w}$ is a saturation function such that the element in $\mathbf{w}(k+i)$ is upper bounded as $w(k+j) \leq \Psi_{\max} < \infty, j \in \mathbb{N}_{[0,i-1]}$. Different from previous control policy design, the saturation function $\Psi_i(\cdot)$ renders the policy nonlinear. As shown in [73], the recursive feasibility and mean-square stability can be established by this nonlinear control policy even in presence of hard input constraints and unbounded uncertainties. Extensions of this type of nonlinear control policy include output feedback implementation [68] and vector-space approach [74].

(A_3) *Stochastic programming or scenario-based approach.*

The first attempt of using stochastic programming techniques to solve the stochastic OCP (1.8) is reported in [75] for linear systems subject to additive disturbances. An optimization scenario tree is generated based on a maximum-likelihood method, and a multi-stage stochastic optimization problem is formulated. To consider probabilistic state constraints, an extension of this approach is reported in [47]. Since the number of nodes in the optimization tree grows exponentially as the prediction horizon increases, evaluation of chance constraints on every node is computationally intractable in general.

To obtain a tractable surrogate for the stochastic programming problem, the randomized approach or sample-based approach [76] has been utilized. In the sample-based approach, a sufficient amount of uncertainty realizations will be generated online, and accordingly, a suitable approximation to the stochastic OCP (1.8) can be obtained with these realizations. Compared with other SMPC approaches, this method can be utilized in more general situations at the price of a heavier online computational load. In [41], a sample-based SMPC algorithm is proposed for linear Markov jumping systems with arbitrary disturbance distributions. A property in [41] shows that as the number of samples approaches to infinity, an exact deterministic OCP can be obtained to approximate the stochastic OCP (1.8). Extensions of this approach include [47] for Markovian-jump systems and [77] for Markovian-switching systems. It should be noted that no theoretical guarantee on the number of samples at each time instant is provided in the aforementioned sampled-based SMPC methods. This leads to a drawback that sample-based SMPC algorithms are generally computationally intensive, implying that they are impractical for implementation, especially for large scale nonlinear systems.

Significant improvement to tackling the drawback in sample-based stochastic programming has been proposed in the scenario-based approach [78], [61]. In the scenario-based approach, instead of verifying original chance constraints in (1.7), a finite amount of sampled deterministic constraints is verified. The sampled constraints are selected by sampling the original chance constraints, and an explicit bound on the size of generated scenarios is given to guarantee the probability level of constraint violation. In [62] and [51], scenario-based SMPC algorithms are developed for linear systems subject to multiplicative uncertainties and additive disturbances. For chance constraints in form of $\mathrm{Pr}_k\{g^\mathrm{T}x(i|k) \le h\} \ge 1 - p$, the scenario-based reformulation is given as $g^\mathrm{T}(\bar{x}(i|k) + E_i\mathbf{w}^{[i,k_i]}(k)) \le h,$, where $k_i \in \mathbb{N}_{[1,N_{s,i}]}$ is the samples index for time step $k + i$, $\mathbf{w}^{[i,k_i]}(k)$ denotes the k_ith disturbance realizations and $N_{s,i}$ is the total number of samples at time $k + i$. Hence, from the results in [78], the constraint satisfaction probability $\mathrm{Pr}_k\{g^\mathrm{T}x(i|k) \le h\} \ge 1 - p$ with a confidence level β is given by

$$\sum_{k=1}^{d_i-1} \binom{N_{s,i}}{k} p^k (1-p)^{N_{s,i}-k} \geq \beta,$$

where d_i is the number of variables in the optimization problem. The resulting OCP is formulated as a convex deterministic cost function subject to hard input and sampled state constraints. The number of required scenarios $N_{s,i}$ is selected according to the sufficient bound provided in [78] as

$$N_{s,i} \geq \frac{d_i + 1 + \ln(1/\beta) + \sqrt{2(d_i + 1)\ln(1/\beta)}}{p}.$$

It should be noted that the sufficient bound $N_{s,i}$ as suggested in [78] is usually conservative since the number of extracted scenarios is more extensive than what is needed. To deal with this problem, an iterative sample-removal algorithm is proposed in [61]. In [79], to reduce the size of generated samples, only part of the sampled set is utilized to construct chance constraint reformulation. Furthermore in [52], the average-in-time chance constraints are considered instead of the pointwise-in-time form. The support rank of chance constraints determines the bound $N_{s,i}$, which is independent of the state dimension, for the required amount of scenarios. This property significantly reduces the size of the sample set even for large scale systems. The method is then further extended in [80], where an improved lower bound is achieved when chance constraints have certain structural properties. In scenario-based SMPC, closed-loop theoretical properties such as recursive feasibility and stability are not well established in general. In a recent work [81], the online sampling of uncertainties is replaced by an offline sampling scheme, and the asymptotical stability of the chance-constrained system can be established.

(A$_4$) *Nonlinear SMPC (NSMPC) approach.*

The development of SMPC algorithms for nonlinear systems has drawn little attention in the literature. The primary challenge in designing NSMPC algorithms lies in the lack of efficient uncertainty propagation methods through nonlinear dynamics. Different from linear systems, the state decomposition as in (1.9) cannot be conducted since nonlinear systems do not have the superposition property. Based on different uncertainty propagation methods, NSMPC can be categorized into the following methods: generalized polynomial chaos expansions (gPCEs) approach [34, 82], Gaussian-mixture (GM) approximation approach [83, 84, 70, 71, 85], and Fokker-Planck (FP) equation approach [35]. In gPCEs based approach, polynomial chaos expansions are utilized to obtain a surrogate for the nonlinear dynamics, which provides an efficient way to predict the state evolution. The nonlinear model (1.1) is approximated by expansions of orthogonal polynomial basis functions, and the statistical moments of predicted state can be computed from expansion coefficients. In this way, the chance constraint $\mathrm{Pr}_k\{g^\mathrm{T}x(i|k) \leq h\} \geq 1 - p$ can be readily reformulated as a second-order cone expression. Gaussian-mixture approximation SMPC approach relies on its universal approximation property to predict the probability distribution of stochastic state variables along with the disturbed nonlinear dynamics. Different from gPCEs based approach, where the cost function is defined in terms of some specific moments of states, the cost function in GM based approach utilizes the complete probability distribution of state variables. Similarly, the Fokkler-Plank equation is utilized in [35] to describe the state evolution for nonlinear input affine system subject to probabilistic joint constraint. Two common shortcomings of the aforementioned NSMPC approaches are: a) The computational complexities of uncertainty propagation methods are intensive; b) establishing the closed-loop theoretical properties for NSMPC algorithms is still a challenging problem.

In the context of MPC, the concept of recursive feasibility means that if the MPC problem can be solved for the initial state $x(k)$, then the MPC problem is feasible for any subsequent states $x(k + i), i \in \mathbb{N}_{\geq 1}$. However, it is not easy to guarantee such property for the aforementioned SMPC algorithms. We consider linear systems subject to unbounded disturbances as an example. Suppose the stochastic OCP (1.8) is feasible at the last time instant $k - 1$. At the time instant k, the initial nominal predicted state is $\bar{x}(0|k) = x(k) = f(x(k - 1), u^*(k - 1), w(k - 1))$, implying that the initial predicted state $\bar{x}(0|k)$ contains the realization of uncertainties $w(k - 1)$. If the uncertainty at the last time instant $w(k - 1)$ takes unbounded large values (even with low probability), then it may be impossible for predicted states $\bar{x}(i|k), i \in \mathbb{N}_{[1,N-1]}$ to satisfy the constraint $\mathrm{Pr}_k\{g^\mathrm{T}x(i|k) \leq h\} \geq 1-p$. Since the feasibility guarantee of the associated optimization problems directly determines the successful implementation of the SMPC algorithms, it is of paramount importance to rigorously establish conditions to ensure feasibility. In the literature, three main approaches on the feasibility analysis have been proposed for **linear systems**:

1. *Recursive feasibility guarantee with a probability.* For example, as proved in [59], if the MPC problem is feasible at time instant k, then it will be feasible for the subsequent states $x(k+1), \ldots, x(k+N)$ only with a given probability. In stochastic tube approaches [48, 45], the ellipsoidal invariant tube with probability is constructed by extending the invariant analysis to stochastic systems. When the optimization problem is infeasible, an alternative feasible optimization problem is solved to steer the state back to the invariant set.

2. *Strict recursive feasibility guarantee.* Based on the assumption that the system is subject to bounded uncertainty, a mixed probabilistic/worst-case constraints tightening technique [46, 43, 69] is designed in stochastic tube approach. The chance constraint $\mathrm{Pr}_k\{g^\mathrm{T} x(i|k) \leq h\} \geq 1 - p$ is only imposed at time $i = 1$, and the worst-case realizations of $w(k+i)$ are considered for $i \in \mathbb{N}_{[2,N-1]}$, implying robust constraints tightening over subsequent prediction horizon.

3. *Recursive feasibility guarantee with initial constraint.* In [56, 72], besides the original initial condition, the pair $\{\bar{x}(0|k) = \bar{x}(0|k-1), \mathrm{Var}[e(0|k)] = \mathrm{Var}[e(1|k-1)]\}$ is imposed to the SMPC problem as an additional initial constraint to guarantee the recursive feasibility. Define $\bar{x}(0|k-1) = \mathbb{E}_{k-1}\{x(1|k-1)\}$ and $\mathrm{Var}[e(1|k-1)] = \mathbb{E}_{k-1}[(x(0|k)-\mathbb{E}_{k-1}[x(1|k-1)])(x(0|k)-\mathbb{E}_{k-1}[x(1|k-1)])^\mathrm{T}] \neq 0$.

Stability analysis of the resulting closed-loop system is another important task. In the literature, the mean-square stability is generally used. If considering additive uncertainties, it can be proved that

$$\lim_{k\to\infty} \mathbb{E}[\|x(k)\|^2] = \lim_{k\to\infty} (\|\mathbb{E}[x(k)]\|^2 + \mathrm{var}(x(k))) \leq c,$$

where c is a constant. This implies that the system state is steered to the neighborhood of origin. When considering multiplicative uncertainties, the point-wise convergence to the origin can be achieved as shown in [48].

It should be noted that the recursive feasibility of the SMPC problem in stochastic programming approach and NSMPC approach is usually given by assumption. This poses theoretical challenges for these two approaches. Meanwhile, few results in NSMPC consider the stability of the closed-loop system, except for the pioneering work [86] where Lyapunov constraints are introduced in the stochastic OCP. A summary of recent progress in the SMPC stability and convergence analysis is given in [87].

1.1.3 Distributed stochastic MPC

With the development of computer technologies and communication networks, cyber-physical systems (CPSs) have become an interest of research due to the comprehensive integration of physically engineered systems, such as sensors, actuators and plants, with intricate cyber components, possessing information communication and computation. In CPSs, advantages of low installation cost, high reliability, flexible modularity, improved efficiency, and greater autonomy can be obtained by the tight coordination of physical and cyber components. Several sectors, including robotics, transportation, health care, smart buildings, and smart grid, have witnessed the successful application of CPSs design. Due to the heterogeneous and spatially interconnected nature in CPSs, it necessitates the adoption of a distributed control structure to improve the structural flexibility and scalability requirements while maintaining some desirable closed-loop properties. Meanwhile, the integration of extensive cyber capability and physical plants with ubiquitous uncertainties also introduces concerns over the robustness and stability of the CPSs. Thus, in order to achieve satisfactory performance metrics of efficiency, robustness and stability, a comprehensive investigation into control synthesis of CPSs under the distributed paradigm is of importance. The distributed model predictive control (DMPC) synthesis plays a vital role in CPSs design since the general nonlinear dynamics and state or input constraints can be systematically handled under this framework. In the DMPC framework, the overall system is divided into many subsystems, and subsystems can communicate with each other. Then a local MPC controller for each subsystem can be designed to meet the requirement for each subsystem. For a local MPC controller, the predicted control actions for actuators and the corresponding state trajectories of the subsystem can be easily obtained. Then, the predicted information can be transmitted to the neighbour subsystems. This feature can help to improve the control performance of the overall system.

The research on distributed SMPC (DSMPC) for large scale interacting systems subject to probabilistic uncertainties has drawn increasing interest in the last decade. Pioneering works in the area include [88], where the state-affine based SMPC algorithms [56, 72] are applied to distributed decoupled system with coupled chance constraints. For linear decoupled systems subject to additive disturbances and coupled chance constraints, DSMPC methods have been proposed in [89] using gPCEs-based technique and in [90] using stochastic tube technique. Mean square stability of the closed-loop surrogate system has been shown in these two works [89, 90]. For unbounded additive disturbances, DSMPC algorithms for linear systems subject to hard input constraints are developed in [91, 92]. DSMPC algorithms for systems with bounded or unbounded multiplicative uncertainties have been investigated in [93, 94]. Recently, based on the stochastic tube approach, [95] studies the DSMPC algorithm for linear systems with both multiplicative and additive uncertainties. The scenario-based SMPC approaches have also been extended to distributed MPC area, as shown in [96, 97]. Considering the output-feedback control parameterization, the output-feedback based DSMPC approaches are studied in [89, 98]. Moreover, in [91], DSMPC tracking control for linear systems subject to coupled constraints is proposed.

Table 1.5: Distributed stochastic MPC review summary

	SMPC method	
	Stochastic tube	Scenario-based
Uncertainty assumptions		
Add. & Bounded	[89, 90]	
Add. & Unbounded	[91, 92]	[97, 98]
Multi. & Bounded	[93]	
Multi. & Unbounded	[94]	
Add. & Multi. & Bounded	[95]	[96]

In summary, developments in the area of DSMPC are given in Table 1.5. It should be noted that existing DSMPC methods rely on extending the well-developed uncertainty propagation methods in SMPC to the diagram setup. Meanwhile, the communication protocol among subsystem utilizes the sequential update rule as suggested in [99]. In general, a fully connected communication topology is assumed, and network issues such as data dropouts or time delays are not considered in existing works. More challenging issues such as advanced system decomposition method for large-scale stochastic systems, more efficient communication protocols design among subsystems or advanced stochastic distributed optimization algorithms are still open.

1.2 Literature review on event-based MPC

In recent years, with the increasing development of CPSs, the theory on aperiodic control or event-based control has been developed significantly due to two reasons [100]. The first reason is to handle the cost, computation, and communication constraints in CPSs explicitly. The second one is that some benefits of the event-based control can be introduced in [101, 102]. In general, event-based control can be classified into event-triggered control and self-triggered control. Motivated by the advantages proposed in [101], several important papers [103, 104, 105] with systematical design of the stabilizing event-triggered controller for linear or nonlinear systems are proposed, respectively. At the same time, based on the aperiodic sampling scheme, another approach [106] called self-triggered control is introduced for the first time.

Two elements are of importance in the event-triggered and self-triggered control systems. The first one is a feedback controller that computes the control input. Different control methods, such as PID control, optimal control, nonlinear control, or other types of control methods with specific system requirements, can be utilized to design the corresponding controller. The second one is a triggering mechanism that determines when the control inputs are updated. The main difference between event-triggered control and self-triggered control is the triggering mechanism. In the event-triggered control, a trigger will be generated when the continuously measured system state violates a preset threshold. In the self-triggered control, the next sampling time instant will be computed at the current sampling time instant based on the system dynamics and current state.

1.2.1 Self-triggered SMPC

Currently, it is a tendency to take constraints, such as communication constraints, computational constraints, and input/ state constraints, into account explicitly in the design of feedback control law. Also, for CPSs design, it is crucial to consider the event-triggered, and self-triggered implementation of the control law since the computational and communication load of the control system can be reduced. Meanwhile, MPC is an ideal candidate for constraint handling and multivariable control purposes, and many research efforts have been made in combining the MPC with the event-based control method. In [107], an event-triggered predictive control strategy is proposed, and the triggering condition is given by comparing the system state and the forecast states continuously. Once the difference is greater than a preset threshold, an event is triggered, and a new control input is updated by the controller. In [108], an event-triggered MPC algorithm based on the output-feedback is proposed for linear systems. In [109], an event-triggered MPC for nonlinear continuous-time systems has been proposed, and the triggering condition is designed based on the concept of input-state stability.

The integration of SMPC and self-triggered MPC can ensure closed-loop chance constraints satisfaction, and therefore reduce the inherent conservativeness of RMPC. Alternative stochastic self-triggered MPC schemes are available in the literature, such as [110, 111, 112]. In [110] and [111], the similar self-triggering condition inspired by [113] is adopted, while different chance constraints handling methods from [43] and [72] are utilized, respectively. In [110], the stability is analyzed in the mean-square sense, whereas the input-state stability of the closed-loop system is proved in [111]. In [112], the self-triggering condition is designed based on the summation of the MPC value function bound and the performance measure at the last sampling time. One common feature of previously mentioned works on self-triggered SMPC is that the open-loop control paradigm is applied between triggering time instants, and resulting constraint tightening parameters are therefore more complex and time-varying compared to conventional SMPC methods. The triggering condition design relies on the bounds of the MPC value function, where a periodical sampling is assumed after the open-loop phase. In addition, to evaluate the triggering condition at each sampling time instant, the solution to a set of quadratic programs with time-varying tightened constraints is required.

Table 1.6: Self-triggered MPC review summary.

	Uncertainty	Linear	Nonlinear
STMPC	Robust	[113]	[114]
ETMPC	Robust	[115]	[116]
OUPUT ET/STMPC	Robust	[117]	[118]
ET/STMPC	Stochastic	-	[111, 110, 112]

1.3   Motivation and organization of Ph.D. book

In previous sections, we have reviewed the development of stochastic MPC, the evolution of distributed SMPC and the growth of event-based MPC. Specifically, we have distinguished the difference between stochastic MPC and robust MPC, demonstrated the improvement of distributed SMPC and clarified the diversities between event-triggered MPC and self-triggered MPC. Based on the understanding of these various technologies and application backgrounds, we could establish a novel control framework for large scale CPSs where chance constraints and the communicational load among subsystems should be considered explicitly. The model uncertainties and communication uncertainties can be handled by stochastic MPC; the cooperation between each subsystem can be handled by distributed algorithm; and the communication load can be reduced by the self-triggered mechanism. The integration of these technologies will lead to novel control methods that can be applied in many areas, such as industrial process control, smart grid and autonomous vehicle.

Chapter 1 provides an overview of the current development of stochastic MPC, distributed SMPC and event-based MPC. For each topic, some representative methods have been demonstrated. Existing problems and promising directions are introduced as well.

Chapter 2 presents a stochastic self-triggered model predictive control scheme for linear systems with additive uncertainty and with the states and inputs being subject to chance constraints. In the proposed control scheme, the succeeding sampling time instant and current control inputs are computed online by solving a formulated optimization problem. The chance constraints are reformulated into a deterministic fashion by leveraging the Cantelli's inequality. Under few mild assumptions, the online computational complexity of the proposed control scheme is similar to that of a nominal self-triggered MPC algorithm. Furthermore, initial constraints are incorporated into the MPC problem to guarantee the recursive feasibility of the scheme, and the stability conditions of the system have been developed. Finally, numerical examples are provided to illustrate the achievable performance of the proposed control strategy.

Chapter 3 discusses a stochastic self-triggered model predictive control algorithm with adaptive prediction horizon for linear systems subject to additive uncertainties and state chance constraints. The communication cost is explicitly considered by adding a damping factor in the cost function. A novel self-triggered condition is proposed and the asymptotic sampling behaviour is analyzed. Sufficient conditions are provided to guarantee closed-loop chance constraints satisfactions. Furthermore, the recursive feasibility of the algorithm is analyzed, and the closed-loop system is shown to be quadratically stable. Finally, the effectiveness of the control method is verified by numerical examples.

Chapter 4 proposes a distributed self-triggered stochastic MPC control scheme for CPSs under coupled chance constraints and additive disturbances. To mitigate performance degradation due to the implementation of self-triggered mechanism, a self-triggered MPC optimization problem is defined. Both the next sampling time instant and resulting control action sequences are determined by solving the self-triggered problem and then transmitted from controller to actuator through communication networks at each sampling time instant. Based on the information on stochastic disturbances, both local and coupled chance constraints are transformed into the deterministic form using the tube-based method. Improved terminal constraints are constructed to guarantee the recursive feasibility of the control scheme. Theoretical analysis has shown that the overall closed-loop CPSs are quadratically stable. Numerical examples illustrate the efficacy of the proposed control method in terms of data transmission reduction.

Chapter 5 concludes the book and suggests some promising directions for future research.

Chapter 2

Stochastic Self-triggered MPC for Linear Constrained Systems under Additive Uncertainty and Chance Constraints

2.1 Introduction

Stochastic model predictive control (MPC) has received considerable attention because it is capable of optimizing the system performance under stochastic uncertainties and chance constraints on the state and input variables. The development of stochastic MPC has stimulated a wide range of applications in industry, such as building climate control [5, 7] and automotive control [119, 120]. In contrast to robust MPC, which relies on the worst-case consideration on the uncertainties, stochastic MPC makes use of the information about the distribution of the uncertainties. If the uncertainties are characterized as random processes, it is desirable to reformulate the constraints in a probabilistic framework. Also, stochastic MPC caters for many cases in which the constraints are probabilistic in nature. As shown in [81], for the same prediction horizon, if the constraints are formulated as chance constraints, the region of attraction will be enlarged significantly.

Two cruxes exist in the design of a stochastic MPC algorithm: (i) reformulating chance constraints into deterministic representations and (ii) theoretically analyzing stability and recursive feasibility. As stated in [57] two main approaches to the former have been proposed to handle the optimization problem with chance constraints: analytical approximation methods and scenario-based methods. For linear systems subject to additive uncertainties, various methods cast the stochastic optimal control problems with chance constraints as a tractable problem by deterministic reformulations of the chance constraints [48, 69, 43, 121]. The online computational complexity of the resulting algorithm is comparable to that of a nominal MPC, but some degree of conservativeness is introduced due to the approximation of the chance constraints. Alternatively, in scenario-based methods [52, 61, 81], a set of disturbance realizations is randomly generated to find the optimal solution of the stochastic MPC problem with arbitrarily high accuracy. It is worth noting that scenario-based methods cope with generic probability distributions, cost functions and chance constraints. In comparison with analytical methods, the resulting algorithm is computationally demanding because a large number of disturbance realizations is required for the online computation.

As discussed above, the stochastic MPC schemes are executed periodically on digital platforms. In networked control systems (NCSs), whose components are connected through a communication network, the communication cost among components cannot be neglected, and the high communication load is the main concern for implementing stochastic MPC. If the components are connected through wireless networks, the communication load will be heavier, possibly leading to packet dropouts and network-induced delays. These challenges, introduced by the communication network, may degrade the system performance and even destabilize the control system [122], [123]. To deal with these challenges, the aperiodic sampling scheme is a promising solution since a considerable amount of communication load can be reduced. For consensus problems in the multi-agent system in which agents share information through the networks, a novel event-triggered transmission strategy is reported in [124]. Reviews on NCSs considering the aperiodical control and filtering schemes are referred to [125, 126, 127, 128].

In periodic sampling schemes, without considering the particular dynamics of the system, this general implementation can lead to redundant samplings. However, in aperiodic sampling schemes, the control inputs are updated only when the system performance cannot meet some specified requirements (i.e., the performance index violates some predefined thresholds), and this sampling mechanism can lead to a lower average sampling rate. Results in [129, 100] have highlighted these advantages, and since then, several results on MPC using aperiodic sampling schemes have been proposed. Recently, robust event-triggered MPC and robust self-triggered MPC have been proposed, and these control strategies have received increasing attention. For nonlinear continuous-time systems affected by additive uncertainties, an event-triggered robust MPC algorithm has been proposed in [116]. For nonlinear input-affine dynamical systems, a self-triggered MPC control scheme, in which the control sequence is adaptively sampled is reported in [114]. For linear systems, the co-design problem of jointly determining the control input to the plant and the next sampling instant has been discussed in [130, 131, 110]. The authors in [132, 113] separate the problem to a bilevel optimization problem while tube-based MPC is utilized to deal with the additive disturbance.

Note that the aforementioned self-triggered MPC algorithms only consider hard constraints, with an exception in a most recent work [110], where the chance constraints are considered. In this study, we aim to develop a stochastic MPC algorithm for disturbed linear systems under the framework of *self-triggered mechanism*. The difference between our proposed work and the existing work is that we take the possibly unbounded stochastic uncertainty and chance constraints into account. The difference between SMPC and RMPC makes our work essentially different from the existing work on robust self-triggered MPC. As a result, the main challenges of this work are: How to propagate the uncertainties during two sampling instants; and how to formulate a tractable optimization problem in the presence of chance constraints. Comparing with [110], chance constraints in our work are reformulated in a completely different way; consequently, the resulting theoretical analysis is inherently different. The main contribution of this work is two-fold:

- A stochastic self-triggered MPC scheme is proposed for linear systems under additive uncertainty and chance constraints. The chance constraints on the states and inputs are reformulated into deterministic terms by leveraging the Cantelli's inequality [56, 72]. At each sampling time instant, the co-design problem of deciding the next sampling instant and the control input sequence during the inter-execution time interval is addressed by solving a set of optimization problems. With the proposed aperiodic scheduling strategy, the controller only needs to sample the state and transmit control input when necessary, therefore reducing the communication load between the sensor and controller significantly.

- Theoretical analysis of the proposed stochastic self-triggered MPC algorithm is performed. Tightened constraints on the state and control input are designed to guarantee the satisfaction of chance constraints of the proposed control scheme. In [110], the recursive feasibility is guaranteed based on the assumption that the uncertainty is bounded, while in our method, additional initial constraints are imposed to ensure the recursive feasibility of the scheme. Meanwhile, sufficient conditions under which the closed-loop system is stable are given, and it has been shown that the system state will converge to an invariant set around the origin.

The remainder of this chapter is organized as follows. Section 2.2 introduces the formal problem formulation of the work, where the reformulation of chance constraints and constructions of constraint sets are presented. In Section 2.3, the proposed stochastic self-triggered MPC problem is defined. Following that, the closed-loop properties of the proposed control scheme are summarized in Section 2.4, and sufficient conditions to guarantee the stability of the system are given. In Section 2.5, the advantages of the proposed control scheme are demonstrated by numerical examples. Section 2.6 concludes this chapter.

Notations: $\mathbb{N}$ denotes the set of integers, and $\mathbb{N}_{[a,b]}$ represents the set of integers from a to b, where $a \leq b, a, b \in \mathbb{N}$. $\mathbb{R}^n$ stands for the n-dimensional real space. $\mathbb{E}\{\cdot\}$ and $\text{var}\{\cdot\}$ denote the expectation and variance of a random variable, respectively. For a matrix X, X^{T} denotes the transpose of X, and $\text{tr}(X)$ denotes the trace of X. $X = \text{diag}(x_1, x_2, \ldots, x_n)$ denotes a diagonal matrix with elements $x_1, x_2, \ldots, x_n$. The maximum and minimum eigenvalues of X are denoted by $\bar{\lambda}(X)$ and $\underline{\lambda}(X)$, respectively. Given a set $\mathcal{B}$ and a point η, $d(\eta, \mathcal{B}) := \inf\{\|\eta - b\|, b \in \mathcal{B}\}$ denotes the point-to-set distance. $\mathcal{B}_r := \{x \in \mathbb{R}^n : \|x\| \leq r\}$ denotes the ball with a radius of r around the origin.

2.2 Problem formulation

We consider the following discrete-time linear system

$$x(k+1) = Ax(k) + Bu(k) + Fw(k), k \in \mathbb{N}_{\geq 0}, \tag{2.1}$$

where $x(k) \in \mathbb{R}^{n_x}$ is the system state, $u(k) \in \mathbb{R}^{n_u}$ is the control input, and $w(k) \in \mathbb{R}^{n_w}$ is an additive uncertainty with zero mean and a known variance W, and possibly unbounded support. It is assumed that $w(k)$ is independent and identically distributed.

Assumption 1. *The state is perfectly known at each time instant, and the pair (A, B) is stabilizable.*

The state and input variables are subject to the following single chance constraint: For time instant k, and $i \in \mathbb{N}_{\geq 0}$

$$P\{b_r^{\mathrm{T}} x(k + i) \geq 1\} \leq p_{x,r}, r \in \mathbb{N}_{[1,n_r]}, \tag{2.2}$$

$$P\{c_s^{\mathrm{T}} u(k + i) \geq 1\} \leq p_{u,s}, s \in \mathbb{N}_{[1,n_s]}, \tag{2.3}$$

where $P\{\cdot\}$ denotes the probability of the constraint violation, $b_r, r \in \mathbb{N}_{[1,n_r]}$ and $c_s, s \in \mathbb{N}_{[1,n_s]}$ are constant vectors, and $p_{x,r}, p_{u,s}$ are design parameters for each single chance constraint. It is assumed that both polyhedrons defined by $b_r^{\mathrm{T}} x(k) \leq 1$ and $c_s^{\mathrm{T}} x(k) \leq 1$ contain the origin.

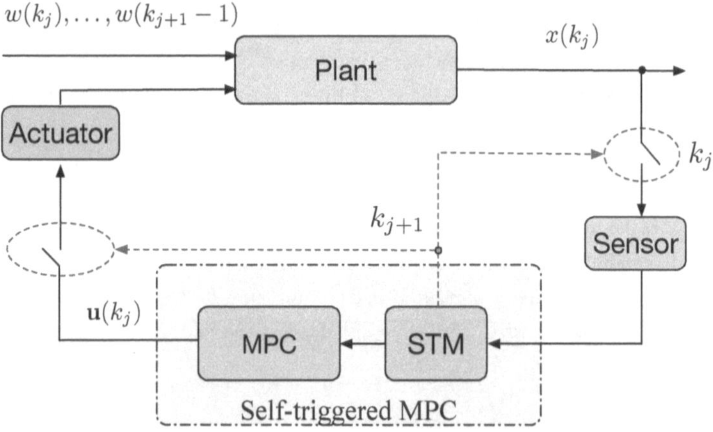

Figure 2.1: The schematic diagram of the stochastic self-triggered MPC system.

The system framework is shown in Figure 2.1. To reduce the frequency of information transmission from the sensor to the controller, a self-triggered mechanism (STM) is introduced to the framework of stochastic MPC. The triggering time sequence is defined as $\{k_j\}, j \in \mathbb{N}_{\geq 0}$. At sampling time instant k_j, the next sampling time instant k_{j+1} and the control input sequence $\mathbf{u}(k_j) = \{u(k_j + i), \ldots, u(k_{j+1} - 1)\}, i \in$

$\mathbb{N}_{[0,k_{j+1}-k_j-1]}$ are jointly determined by a self-triggered controller as shown below:

$$u(k_j + i) = u^*(i|k_j), i \in \mathbb{N}_{[0,k_{j+1}-k_j-1]}, \tag{2.4a}$$

$$k_{j+1} = k_j + l^*(k_j), j \in \mathbb{N}_{\geq 0}, k_0 = 0, \tag{2.4b}$$

where $\bar{\mathbf{u}}^*(l^*(k_j)) = \{u^*(0|k_j), \ldots, u^*(l^*(k_j) - 1|k_j)\}$ is the optimal control input sequence and $l^*(k_j)$ is the optimal inter-execution time interval. $\bar{\mathbf{u}}^*(l^*(k_j))$ and $l^*(k_j)$ are determined by solving the stochastic self-triggered optimal control problem defined in the next section. Then, the controller sends the obtained control sequence to the actuator to control the plant. During the time interval $k_{j+1} - k_j$, the system is controlled in an open-loop fashion since no state measurement is required, and the sensors can be turned off. At the next sampling time instant k_{j+1}, this procedure is repeated, resulting in a closed-loop system

$$x(k_j + i + 1) = Ax(k_j + i) + Bu^*(i|k_j) + Fw(k_j + i), i \in \mathbb{N}_{[0,k_{j+1}-k_j-1]}, j \in \mathbb{N}_{\geq 0}. \tag{2.5}$$

Let $\bar{x}(k_j + i) = \mathbb{E}\{x(k_j + i)\}$ denote the nominal state of $x(k_j + i)$. Since $w(k)$ is assumed to be a disturbance with zero-mean, the predicted nominal state $\bar{x}(i|k_j)$, given the state $\bar{x}(0|k_j)$ at time instant k_j, evolves according to

$$\bar{x}(i + 1|k_j) = A\bar{x}(i|k_j) + B\bar{u}(i|k_j), i \in \mathbb{N}_{[0,N-1]}, \tag{2.6}$$

in which $\bar{u}(i|k_j)$ is the predicted nominal control input, and it is determined online by solving the optimization problem defined in the next section. Consider the state-feedback control law at time instant k_j,

$$u(i|k_j) = \begin{cases} \bar{u}(i|k_j), & i \in \mathbb{N}_{[0,l-1]}; \\ \bar{u}(i|k_j) + K_{k_j+i}(x(k_j + i) - \bar{x}(i|k_j)), & i \in \mathbb{N}_{[l,N-1]}, \end{cases} \tag{2.7}$$

in which the predicted input and gain sequences $\bar{\mathbf{u}}_N = \{\bar{u}(0|k_j), \ldots, \bar{u}(N - 1|k_j)\}$, $\mathbf{K}_{N-l} = \{K_{l|k_j}, \ldots, K_{N-1|k_j}\}$ are defined as the result of a suitable optimization problem solved at time instant k_j. $l \in \mathbb{N}_{[1,N-1]}$ is defined as the inter-execution time. In the first $l - 1$ steps, the system is controlled in an open-loop fashion, and the state measurements are not required in this period. The state feedback is introduced in the last $N - l$ steps to deal with the disturbance. The open-loop controlled phase is involved because of the self-triggered mechanism. As pointed out in [2], when the system is affected by disturbances, closed-loop MPC shows great advantages compared with open-loop MPC. This type of control law (2.7) is also used in a recent work [113].

Define $e(i|k_j) = x(k_j + i) - \bar{x}(i|k_j)$ for $i \in \mathbb{N}_{\geq 0}$, as the error between the real state and the predicted nominal state. And $E(i|k_j) = \text{var}\{e(i|k_j)\} = \mathbb{E}\{e(i|k_j)e(i|k_j)^\mathrm{T}\}$ is defined as the variance of the predicted error term. It can be shown that $E(i|k_j)$ evolves according to the following equations

$$E(i|k_j) = \begin{cases} A^i E(0|k_j)(A^i)^\mathrm{T} + \sum_{h=0}^{i-1} A^h FW(A^h F)^\mathrm{T}, & i \in \mathbb{N}_{[0,l-1]}, \quad (2.8\text{a}) \\[2em] \left[\prod_{h=1}^{i-l} \Phi_{l+h-1|k_j}\right] E(l-1|k_j) \left[\prod_{h=1}^{i-l} \Phi_{l+h-1|k_j}\right]^\mathrm{T} \\ + \sum_{h=1}^{i-1} \Phi_{l+h-1|k_j} FW(\Phi_{l+h-1|k_j} F)^\mathrm{T}, & i \in \mathbb{N}_{[l,N]}, \quad (2.8\text{b}) \end{cases}$$

in which $\Phi_{l+h-1|k_j} = A + BK_{l+h-1|k_j}$ for $i \in \mathbb{N}_{[l,N-1]}$, $h \in \mathbb{N}_{[1,i-1]}$. When the system is sampled periodically, i.e., for $\forall k, l = 1$, the evolution of the predicted variance $E(i|k_j)$ reduces to the one in [56].

Remark 1. *In (2.7), the control sequence $\bar{\mathbf{u}}_N$ is designed to control the predicted nominal state $\bar{x}(i|k)$, for $i \in \mathbb{N}_{[0,N-1]}$; and the feedback gain sequence $\mathbf{K}_{N-l}$ is utilized to restrain the evolution of $E(i|k_j)$, for $i \in \mathbb{N}_{[l,N]}$. At time instant k_j, if the initial condition for the variance is $E(0|k_j) = \underline{0}$, in which $\underline{0}$ is a zero matrix with suitable dimensions, then the evolution of the $E(i|k_j)$ is solely determined by the disturbance variance W in the first $l - 1$ steps.*

2.2.1 Reformulation of the chance constraints

The chance constraints handling is based on the Cantelli's inequality. It has been shown in [56] that the chance constraints can be reformulated as deterministic terms by the following inequality (2.9).

Lemma 1. *[56] Let y be a (scalar) random variable with mean $\bar{y}$ and variance Y. Then for every $0 \leq \alpha \in \mathbb{R}$, it holds that*

$$P(y \geq \bar{y} + \alpha) \leq \frac{Y}{Y + \alpha^2}. \tag{2.9}$$

By using the Cantelli's inequality, the chance constraints (2.2) and (2.3) are verified for time instants $k_j + i, i \in \mathbb{N}_{\geq 0}$, if

$$b_r^{\mathrm{T}} \bar{x}(i|k_j) \leq 1 - \sqrt{b_r^{\mathrm{T}} E(i|k_j) b_r} f(p_{x,r}), r = 1, \ldots, n_r, \tag{2.10}$$

$$c_s^{\mathrm{T}} \bar{u}(i|k_j) \leq 1 - \sqrt{c_s^{\mathrm{T}} U(i|k_j) c_s} f(p_{u,s}), s = 1, \ldots, n_s, \tag{2.11}$$

where $f(p) = \sqrt{(1-p)/p}$, regardless of the specific distribution of the disturbance $w(k)$. The covariance matrix of the control variable $U(i|k_j)$ is defined as

$$U(i|k_j) = \begin{cases} \underline{0}, & i \in \mathbb{N}_{[0,l-1]}; \\ K_{k_j+i}^{\mathrm{T}} E(i|k_j + i) K_{k_j+i}, & i \in \mathbb{N}_{[l,N-1]}, \end{cases}$$

in which $\underline{0}$ is a zero matrix with suitable dimensions. With an additional tightening of the constraint, the state constraints (2.10) and input constraints (2.11) can be linearized as:

$$b_r^{\mathrm{T}} \bar{x}(i|k_j) \leq (1 - 0.5\epsilon) - \frac{1 - p_{x,r}}{2\epsilon p_{x,r}} b_r^{\mathrm{T}} E(i|k_j) b_r, r = 1, \ldots, n_r, \tag{2.12}$$

and for $s = 1, \ldots, n_s$,

$$c_s^{\mathrm{T}} \bar{u}(i|k_j) \leq \begin{cases} 1, & i \in \mathbb{N}_{[0,l-1]}; \\ (1 - 0.5\epsilon) - \dfrac{1 - p_{u,s}}{2\epsilon p_{u,s}} c_s^{\mathrm{T}} U(i|k_j) c_s, & i \in \mathbb{N}_{[l,N-1]}, \end{cases} \tag{2.13}$$

in which $0 < \epsilon < 1$ is a linearization factor.

Remark 2. *Since the Cantelli's inequality is a variant of the Chebyshev's inequality, it can only give a conservative estimate of the original chance constraints. The benefit of replacing the chance constraints by the sufficient upper bound is that the reformulation is obtained without introducing any specific assumptions on the uncertainties. Also, if the uncertainty is assumed to be normally distributed, a less conservative reformulation of the constraint can be achieved.*

2.2.2 Cost function

At sampling time instant k_j, the cost function is defined as

$$
\begin{aligned}
J_l(x(0|k_j), E(0|k_j)) =\mathbb{E}\Bigg\{ &\frac{1}{\alpha}\sum_{i=0}^{l-1}(\|x(k_j+i)\|_Q^2 + \|u(k_j+i)\|_R^2) \\
&+ \sum_{i=l}^{N-1}(\|x(k_j+i)\|_Q^2 + \|u(k_j+i)\|_R^2) + \|x(k_j+N)\|_P^2 \Bigg\},
\end{aligned}
\tag{2.14}
$$

where $\mathbb{E}\{\cdot\}$ denotes the expectation of a random variable. l denotes the inter-execution time interval in the self-triggered control scheme and $\alpha \geq 1$ is a tuning parameter that penalizes the cost in the open-loop phase. N is the prediction horizon. k_j is the current sampling time instant. Q and R are positive definite, symmetric matrices of appropriate dimensions, and P is the solution to the algebraic equation

$$
(A+B\bar{K})^\mathrm{T}P(A+B\bar{K}) + Q + \bar{K}^\mathrm{T}R\bar{K} - P = 0,
\tag{2.15}
$$

where $\bar{K}$ is a suitable stabilizing gain for the pair (A,B). The cost function can be rewritten as $J_l(x(0|k_j), E(0|k_j)) = \bar{J}_l(x(0|k_j)) + \tilde{J}_l(E(0|k_j))$, where

$$
\begin{aligned}
\bar{J}_l(x(0|k_j)) =&\frac{1}{\alpha}\sum_{i=0}^{l-1}(\|\bar{x}(i|k_j)\|_Q^2 + \|\bar{u}(i|k_j)\|_R^2) \\
&+ \sum_{i=l}^{N-1}(\|\bar{x}(i|k_j)\|_Q^2 + \|\bar{u}(i|k_j)\|_R^2) + \|\bar{x}(N|k_j)\|_P^2,
\end{aligned}
\tag{2.16}
$$

is the nominal cost function obtained at k_j, and

$$
\begin{aligned}
\tilde{J}_l(E(0|k_j)) =\ &\frac{1}{\alpha}\sum_{i=0}^{l-1}\mathrm{tr}\{QE(i|k_j)\} \\
&+ \sum_{i=l}^{N-1}\mathrm{tr}\{(Q + K_{k_j+i}^\mathrm{T}RK_{k_j+i})E(i|k_j)\} + \mathrm{tr}\{PE(N|k_j)\},
\end{aligned}
\tag{2.17}
$$

is the predicted variance cost function obtained at k_j. At each sampling time instant k_j, the controller will determine both the optimal inter-execution time interval $l_{k_j}^* \in \mathbb{N}_{[1,N-1]}$, and the corresponding control sequence $\bar{\mathbf{u}}_{l^*(k_j)}^*$ and gain sequence $\mathbf{K}_{N-l^*(k_j)}$.

2.2.3   Terminal constraint

To guarantee recursive feasibility of the algorithm and the stability of the closed-loop system, terminal constraints are enforced at the end of the prediction horizon on both the mean value $\bar{x}(N|k_j)$ and the variance $E(N|k_j)$ as follows,

$$\bar{x}(N|k_j) \in \bar{\mathbb{X}}_f, \tag{2.18}$$

$$E(N|k_j) \leq \bar{E}. \tag{2.19}$$

The set $\bar{\mathbb{X}}_f$ is a positively invariant set for the system with the control law $u(k+i) = \bar{K}\bar{x}(i|k)$, that is

$$(A + B\bar{K})\bar{x}(i|k) \in \bar{\mathbb{X}}_f, \forall \bar{x}(i|k) \in \bar{\mathbb{X}}_f.$$

The dynamics of $E(i|k_j)$ is given in (2.8), and the terminal constraint for $E(N|k_j)$ is the steady state solution of the Lyapunov equation in (2.8b) with a constant stabilizing $\bar{K}$. So $\bar{E}$ verifies the Lyapunov-type equation

$$\bar{E} = (A + B\bar{K})\bar{E}(A + B\bar{K})^{\mathrm{T}} + F\bar{W}F^{\mathrm{T}}, \tag{2.20}$$

where $\bar{W} \geq W$ is an artificially selected matrix for all $\bar{x}(i|k_j) \in \bar{\mathbb{X}}_f$. By choosing the value of $\bar{W}$, we can obtain a larger $\bar{E}$.

The following linearized constraints should also hold for all $\bar{x}(i|k_j) \in \bar{\mathbb{X}}_f$:

$$b_r^{\mathrm{T}}\bar{x}(i|k_j) \leq 1 - \sqrt{b_r^{\mathrm{T}}\bar{E}b_r}f(p_{x,r}), r = 1, \ldots, n_r. \tag{2.21}$$

$$c_s^{\mathrm{T}}\bar{K}\bar{x}(i|k_j) \leq 1 - \sqrt{c_s^{\mathrm{T}}\bar{K}\bar{E}\bar{K}^{\mathrm{T}}c_s}f(p_{u,s}), s = 1, \ldots, n_s. \tag{2.22}$$

2.2.4   Initial constraints

To guarantee the recursive feasibility, the initial conditions $(\bar{x}(0|k_j), E(0|k_j))$ at the current sampling time instant k_j should also be incorporated as free variables in the optimization problem, as shown in the following two strategies.

- Strategy 1: the initial condition for the optimal control problem is chosen as the new state measurement $\bar{x}(0|k_j) = x(k_j)$, and the initial variance $E(0|k_j) = 0$.

- Strategy 2: the initial state and variance variables are set to be the predicted nominal state and variance which are obtained at the last sampling time instant

k_{j-1}, i.e., $\bar{x}(0|k_j) = \bar{x}(k_j|k_{j-1})$, $E(0|k_j) = E(k_j|k_{j-1})$.

The resulting initial constraint for the pair $(\bar{x}(0|k_j), E(0|k_j))$ is given as:

$$(\bar{x}(0|k_j), E(0|k_j)) \in \{(x(k_j), 0), (\bar{x}(k_j|k_j - 1), E(k_j|k_j - 1))\}. \tag{2.23}$$

Remark 3. *In Strategy 1, since the initial condition for $E(0|k_j)$ is reset at every sampling time instant, as stated in Remark 1, $E(i|k_j)$ are all constants for $i \in \mathbb{N}_{[1,N]}$. In this strategy, the predicted variance performance index defined in (2.17) is a constant, and can be removed from the optimization problem.*

2.3 Stochastic self-triggered MPC problem

In this section, we address the stochastic self-triggered MPC synthesis problem. At sampling time instant k_j, by solving the corresponding optimization problem, the controller provides the optimal inter-execution time interval $l^*(k_j)$ and the optimal control sequence $\{u(k_j), \ldots, u(k_{j+1} - 1)\}$ for the interval $[k_j, k_{j+1} + 1), j \in \mathbb{N}_{\geq 0}$ to the plant. With the initial state $\bar{x}(0|k_j)$ and initial variance $E(0|k_j)$, the decision variables of the optimization problem are defined as:

$$\mathbf{d}_l(k_j) = (\bar{x}(0|k_j), E(0|k_j), \bar{u}(0|k_j), \ldots, \bar{u}(N - 1|k_j), K_{l|k_j}, \ldots, K_{N-1|k_j}). \tag{2.24}$$

Given an $l \in \mathbb{N}_{[1,N-1]}$, $\mathcal{D}^l_{k_j}$ is defined as the feasible set:

$$\mathcal{D}^l_{k_j} = \{\mathbf{d}_l(k_j)|(2.6), (2.8), (2.10), (2.11), (2.18), (2.19), (2.23)\}. \tag{2.25}$$

For all $l \in \mathbb{N}_{[1,N-1]}$, define the finite horizon optimization problem as:

$$V_l(\mathbf{d}_l(k_j)) : \min_{\mathbf{d}_l(k_j) \in \mathcal{D}^l_{k_j}} J_l(\mathbf{d}_l(k_j)). \tag{2.26}$$

The stochastic self-triggered MPC problem is defined as:

$$l^*(k_j) := \max\{l \in \mathbb{N}_{[1,L_{\max}]}|\mathcal{D}^1_{k_j} \neq \emptyset, \mathcal{D}^l_{k_j} \neq \emptyset, V_l(\mathbf{d}_l(k_j)) \leq V_1(\mathbf{d}_1(k_j))\}, \tag{2.27a}$$

$$\mathbf{d}^*_{l^*}(k_j) := \underset{\mathbf{d}_{l^*}(k_j) \in \mathcal{D}^{l^*}_{k_j}}{\operatorname{argmin}} J_{l^*}(\mathbf{d}_{l^*}(k_j)), \tag{2.27b}$$

where $L_{\max} \in \mathbb{N}_{[1,N]}$ is the maximal length of the open-loop phase chosen by the designer. For simplicity of presentation, let l^* denote $l^*(k_j)$ in (2.27b). The outer optimization is over the interval length which is an integral multiple of the sampling interval, and the inner optimization is over the control input sequence and the gain sequence. The main idea is to maximize the inter-execution time interval l, while guaranteeing the penalized cost $J_l(x(0|k_j), E(0|k_j))$ defined in (2.14) is not larger than $J_1(x(0|k_j), E(0|k_j))$.

Since (2.10) and (2.11) are nonlinear constraints in terms of decision variables $E(i|k_j)$ and $U(i|k_j)$, the linearization steps in (2.12) and (2.13) can significantly reduce the complexity of the optimization problem. As discussed in [72], we can further reduce the complexity of the optimization problem by using a constant gain $K_{i|k_j} = \bar{K}, i \in \mathbb{N}_{[l,N-1]}$. Under this assumption, the online optimization (2.26) is a standard quadratic programming problem, and $n_u N + n_x + n_x^2$ scalar variables are involved in the optimization problem. Numerically, this will reduce the number of decision variables, and for the finite horizon optimization problem defined in (2.26), the online computational complexity is similar to that of the nominal MPC problem for relatively small-scale problem, e.g., $n_x \leq 10$.

The resulting stochastic self-triggered MPC algorithm is summarized in Algorithm 1:

Algorithm 1: Stochastic self-triggered MPC algorithm

Offline: Determine the linearization factor ϵ, open-loop phase penalizing
parameter α, and the constraints violation probability $p_{x,r}$, $p_{u,s}$.

while *Termination conditions not satisfied* **do**

> **Step 1.** At time instant k_{j+i}, measure the system state $x(k_{j+i})$.
>
> **Step 2.** Solve the optimization problem defined in (2.27). Get the next sampling time instant $k_{j+i+1} = k_{j+i} + l^*(k_{j+i})$, and the control sequence $\{u(k_{j+i}), \ldots, u(k_{j+i+1} - 1)\}$ for this time interval.
>
> **Step 3.** During the time period from k_{j+i} to $k_{j+i+1} - 1$, implement the obtained control sequence $\{u(k_{j+i}), \ldots, u(k_{j+i+1} - 1)\}$.
>
> **Step 4.** Set $k_{j+i} = k_{j+i+1}$, go to Step 1, and repeat the procedure.

end

2.4 Feasibility and stability analysis

In this section, the recursive feasibility of the proposed stochastic self-triggered MPC scheme and the stability of the closed-loop system are analyzed. We first prove that the proposed control scheme is recursively feasible given appropriately chosen terminal constraints on $\bar{x}(N|k_j)$ and $E(N|k_j)$. Then, sufficient conditions are given to guarantee the input-state stability of the closed-loop system.

Theorem 1. *(Recursive feasibility) At the sampling time instant k_j, for any $l \in \mathbb{N}_{[1,L_{max}]}$, $\mathcal{D}^1_{k_{j+1}} \neq \emptyset$, given that the stochastic self-triggered MPC problem admits a solution with decision variables $\mathbf{d}^*_{l^*(k_j)}(k_j) \in \mathcal{D}^{l^*(k_j)}_{k_j}$, and the next sampling time instant is $k_{j+1} = k_j + l^*(k_j)$.*

Proof. At the sampling time instant k_j, by solving (2.27), the stochastic self-triggered MPC problem admits an optimal solution

$$(\bar{x}^*(0|k_j), E^*(0|k_j), \bar{\mathbf{u}}^*_N(k_j), \mathbf{K}^*_{N-l^*(k_j)}(k_j)) \in \mathcal{D}^{l^*(k_j)}_{k_j},$$

where $\bar{\mathbf{u}}^*_N = \{\bar{u}^*_{0|k_j}, \ldots, \bar{u}^*_{N-1|k_j}\}$ is the control sequence, and

$$\mathbf{K}^*_{N-l^*(k_j)} = \{K^*_{l^*(k_j)|k_j}, \ldots, K^*_{N-1|k_j}\}$$

is the gain sequence generated at time instant k_j. At the next sampling time instant $k_{j+1} = k_j + l^*(k_j)$, assume that

$$(\bar{x}^f(0|k_{j+1}), E^f(0|k_{j+1}), \bar{\mathbf{u}}^f_N(k_{j+1}), \mathbf{K}^f_{N-1}(k_{j+1})) \in \mathcal{D}^1_{k_{j+1}},$$

is an admissible but not optimal solution to the stochastic self-triggered MPC problem, in which

$$\bar{x}^f(0|k_{j+1}) = \bar{x}^*(l^*(k_j)|k_j), E^f(0|k_{j+1}) = E^*(l^*(k_j)|k_j),$$
$$\bar{\mathbf{u}}^f_N(k_{j+1}) = \{\bar{u}^*(l^*(k_j)|k_j), \ldots, \bar{u}^*(N-1|k_j), \bar{K}\bar{x}(N|k_j),$$
$$\ldots, \bar{K}(A+B\bar{K})^{l^*(k_j)-1}\bar{x}(N|k_j)\},$$

and

$$\mathbf{K}^f_{N-1}(k_{j+1}) = \{K^*_{l^*(k_j)|k_j}, \ldots, K^*_{N-1|k_j}, \bar{K}, \ldots, \bar{K}(A+B\bar{K})^{l^*(k_j)-1}\}.$$

Note that, $\bar{x}^f(0|k_{j+1})$ and $E^f(0|k_{j+1})$ are chosen according to initial constraints (2.23). Constraints (2.10), (2.11) can be readily verified for $(\bar{x}(k_{j+1}+i|k_j), E(k_{j+1}+i|k_j)), i = 0, \ldots, N - l^*(k_j) - 1$, in view of the feasibility of the ST-SMPC problem at k_j. Constraint (2.18) is verified from the definition of $\bar{\mathbb{X}}_F$. If $\bar{x}(N|k_j) \in \bar{\mathbb{X}}_F$, then

$$\bar{x}(N + i|k) = (A + B\bar{K})^i \bar{x}(N|k_j) \in \bar{\mathbb{X}}_F, \text{ for } i \in \mathbb{N}_{[1,l^*(k_j)]}.$$ From the definition of (2.19), for $i = 1, \ldots, l^*(k_j)$,

$$E(N + i|k_j) = (A + B\bar{K})E(N + i - 1|k_j)(A + B\bar{K})^{\mathrm{T}} + FWF^{\mathrm{T}}$$
$$\leq (A + B\bar{K})\bar{E}(A + B\bar{K})^{\mathrm{T}} + F\bar{W}F^{\mathrm{T}} = \bar{E}.$$

Hence, constraint (2.19) is verified. From (2.21), since $\bar{x}(N + i|k_j) \in \bar{\mathbb{X}}_F$, for $i = 0, \ldots, l^*(k_j)$, we have

$$b_r^{\mathrm{T}} \bar{x}(N + i|k_j) \leq 1 - \sqrt{b_r^{\mathrm{T}} \bar{E} b_r} f(p_x^r) \leq 1 - \sqrt{b_r^{\mathrm{T}} E(N + i|k_j) b_r} f(p_x^r).$$

Therefore, constraint (2.10) is verified. By following the similar lines (2.11) can be verified. The above analyses prove that $\mathcal{D}_{k_{j+1}}^1 \neq \emptyset$. $\qquad\square$

In the following, we provide stability results by presenting Theorem 2.

Theorem 2. *(Stability) If a $\rho \in (0,1)$ exists such that the variance matrix W verifies*

$$\frac{\beta_{max}}{\rho \underline{\lambda}(Q)} \mathrm{tr}(PFWF^{\mathrm{T}}) < \min(\bar{\sigma}^2, \underline{\lambda}(\bar{E})), \tag{2.28}$$

where $\bar{\sigma}$ is the maximum radius of $\mathcal{B}_{\bar{\sigma}}$ included in $\bar{\mathbb{X}}_F$, and β_{max} is defined as

$$\beta_{max} := \max_{l \in \mathbb{N}_{[1,L_{max}]}} N + \frac{\alpha \bar{\lambda}(P)}{\underline{\lambda}(Q)} \left(l - \frac{\Delta(l)}{\mathrm{tr}(PFWF^{\mathrm{T}})} \right), \tag{2.29}$$

in which $\Delta(l) = \frac{1}{\alpha} \sum_{i=0}^{l-1} \mathrm{tr}((l - i - 1)QA^i FW(A^i F)^{\mathrm{T}})$, then

$$d(\mathbb{E}\{\|x(k_j)\|_Q^2\}, [0, \rho^{-1}\beta_{max}\mathrm{tr}(PFWF^T)]) \to 0,$$

as $j \to +\infty$.

Proof. At time instant $k_{j+1} = k_j + l^*(k_j)$, considering the feasible but yet possibly suboptimal control sequence defined in the proof of Theorem 1, and for notation simplicity, we denote the cost at time k_j as $J_{l(k_j)}(k_j)$, for $j \in \mathbb{N}_{\geq 0}$. The predicted cost at the next sampling time instant k_{j+1} is defined as $J_{l(k_{j+1})}(k_{j+1}|k_j)$. From the triggering condition at the next sampling time instant k_{j+1}, the optimal cost computed at k_{j+1} is $J_{l^*(k_{j+1})}^*(k_{j+1}) \leq J_1^*(k_{j+1}) = \bar{J}_1^*(k_{j+1}) + \tilde{J}_1^*(k_{j+1})$. In view of optimality at time k_{j+1},

$$J_1^*(k_{j+1}) \leq J_1^f(k_{j+1}|k_j) = \bar{J}_1(\bar{x}^f(0|k_{j+1}), \mathbf{\bar{u}}_N^f(k_{j+1})) + \tilde{J}_1(E^f(0|k_{j+1}), \mathbf{K}_{N-1}^f(k_{j+1})). \tag{2.30}$$

Note that,

$$\begin{aligned}
\bar{J}_{l^*(k_j)}^*(k_j) - \bar{J}_1^f(k_{j+1}|k_j) &= \frac{1}{\alpha} \sum_{i=0}^{l^*(k_j)-1} (\|\bar{x}(i|k_j)\|_Q^2 + \|\bar{u}(i|k_j)\|_R^2) + \|\bar{x}(N|k_j)\|_P^2 \\
&\quad - \|\bar{x}(N|k_j)\|_Q^2 - \|\bar{K}\bar{x}(N|k_j)\|_R^2 - \|\bar{\Phi}\bar{x}(N|k_j)\|_Q^2 \\
&\quad - \|\bar{K}\bar{\Phi}\bar{x}(N|k_j)\|_R^2 - \cdots - \|\bar{\Phi}^{l^*(k_j)}\bar{x}(N|k_j)\|_P^2 \\
&= \frac{1}{\alpha} \sum_{i=0}^{l^*(k_j)-1} (\|\bar{x}(i|k_j)\|_Q^2 + \|\bar{u}(i|k_j)\|_R^2) - \|\bar{x}(N|k_j)\|_{\bar{P}},
\end{aligned}$$

in which $\bar{P} = -P + (Q + \bar{K}^{\mathrm{T}}RK) + \bar{\Phi}^{\mathrm{T}}(Q + \bar{K}^{\mathrm{T}}RK)\bar{\Phi} + \cdots + (\bar{\Phi}^{l^*(k_j)})^{\mathrm{T}}P\bar{\Phi}^{l^*(k_j)}$. In view of (2.15), we can get $\bar{P} = 0$ by iteration, yielding

$$\bar{J}_{l^*(k_j)}^*(k_j) - \bar{J}_1^f(k_{j+1}|k_j) = \frac{1}{\alpha} \sum_{i=0}^{l^*(k_j)-1} (\|\bar{x}(i|k_j)\|_Q^2 + \|\bar{u}(i|k_j)\|_R^2). \tag{2.31}$$

In addition,

$$\begin{aligned}
\tilde{J}_{l^*(k_j)}^*(k_j) - \tilde{J}_1^f(k_{j+1}|k_j) &= \frac{1}{\alpha} \sum_{i=0}^{l^*(k_j)-1} \mathrm{tr}\{QE(i|k_j)\} + \mathrm{tr}\{PE(N|k_j)\} \\
&\quad - \mathrm{tr}\{(Q + \bar{K}^{\mathrm{T}}R\bar{K})E(N|k_j)\} \\
&\quad - \mathrm{tr}\{(Q + \bar{K}^{\mathrm{T}}R\bar{K})E(N+1|k_j)\} - \cdots - \mathrm{tr}\{PE(N+l^*(k_j)|k_j)\}.
\end{aligned}$$

From (2.8) we have $E(N+i|k_j) = \bar{\Phi}^i E(N|k_j)(\bar{\Phi}^i)^{\mathrm{T}} + \sum_{h=0}^{i-1} \bar{\Phi}^h FW(\bar{\Phi}^h F)^{\mathrm{T}}$ and in view of (2.15), we obtain

$$\tilde{J}_{l^*(k_j)}^*(k_j) - \tilde{J}_1^f(k_{j+1}|k_j) = \frac{1}{\alpha} \sum_{i=0}^{l^*(k_j)-1} \mathrm{tr}\{QE(i|k_j)\} - l^*(k_j)\mathrm{tr}\{PFWF^{\mathrm{T}}\}. \tag{2.32}$$

Combining (2.31) and (2.32), and recalling (2.30), we have

$$J_1^*(k_{j+1}) \leq J_{l^*(k_j)}^*(k_j) - \frac{1}{\alpha} \sum_{i=0}^{l^*(k_j)-1} \mathbb{E}(\|x(k_j+i)\|_Q^2 + \|u(k_j+i)\|_R^2) + l^*(k_j)\mathrm{tr}\{PFWF^\mathrm{T}\}.$$

(2.33)

Since A is assumed to be non-singular, and $Q > 0$, it can be readily verified that $\underline{\lambda}(\sum_{i=1}^{l^*(k_j)}(A^i)^\mathrm{T}QA^i) > 0$, where $\underline{\lambda}(\cdot)$ denotes the minimum eigenvalue of a matrix. It can be shown that

$$\sum_{i=0}^{l^*(k_j)-1} \mathbb{E}(\|x(k_j+i)\|_Q^2 + \|u(k_j+i)\|_R^2) \geq \mathbb{E}\{\|x(k_j)\|_Q^2\} + \sum_{i=1}^{l^*(k_j)-1} \mathrm{tr}\{QE(i|k_j)\}$$

$$= \mathbb{E}\{\|x(k_j)\|_Q^2\} + \sum_{i=1}^{l^*(k_j)-1} \mathrm{tr}\{(A^i)^\mathrm{T}QAE(0|k_j)\}$$

$$+ \sum_{i=0}^{l^*(k_j)-1} \mathrm{tr}\{(l^*(k_j)-i-1)QA^iFW(A^iF)^\mathrm{T}\}$$

$$\geq \mathbb{E}\{\|x(k_j)\|_Q^2\} + \sum_{i=0}^{l^*(k_j)-1} \mathrm{tr}\{(l^*(k_j)-i-1)QA^iFW(A^iF)^\mathrm{T}\}$$

$$\geq \underline{\lambda}(Q)\mathbb{E}\{\|x(k_j)\|^2\} + \sum_{i=0}^{l^*(k_j)-1} \mathrm{tr}\{(l^*(k_j)-i-1)QA^iFW(A^iF)^\mathrm{T}\}.$$

Recalling (2.33), we have

$$J_1^*(k_{j+1}) \leq J_{l^*(k_j)}^*(k_j) - \frac{1}{\alpha}\underline{\lambda}(Q)\mathbb{E}\{\|x(k_j)\|^2\} + l^*(k_j)\mathrm{tr}\{PFWF^\mathrm{T}\}$$

$$- \frac{1}{\alpha} \sum_{i=0}^{l^*(k_j)-1} \mathrm{tr}\{(l^*(k_j)-i-1)QA^iFW(A^iF)^\mathrm{T}\}.$$

(2.34)

Also, from the definition of $J_{l^*(k_j)}(k_j)$, we have

$$J_{l^*(k_j)}(k_j) \geq \frac{1}{\alpha}\mathbb{E}\{\|x(k_j)\|_Q^2\} \geq \frac{1}{\alpha}\underline{\lambda}(Q)\mathbb{E}\{\|x(k_j)\|^2\}.$$

(2.35)

Based on the previous analysis on the cost function, we can proceed to prove the stability of the closed-loop system. Define the terminal set for the nominal state and variance as $\Omega_F = \{(\bar{x}(k), E(k)) : \bar{x}(k) \in \bar{\mathbb{X}}_F, E(k) \leq \bar{E}\}$. For $(\bar{x}(0|k_j), E(0|k_j)) \in \Omega_F$, we can find a corresponding feasible control sequence $\{\bar{K}\bar{x}(0|k_j), \ldots, \bar{K}\bar{\Phi}^{N-1}\bar{x}(0|k_j)\}$. The auxiliary nominal and variance cost functions with respect to the feasible control input sequence can be defined as

$$\bar{J}_1^a(k_j) = \sum_{i=0}^{N-1} \|\bar{\Phi}^i \bar{x}(i|k_j)\|_Q^2 + \|\bar{K}\bar{\Phi}^i \bar{x}(i|k_j)\|_R^2 + \|\bar{\Phi}^N \bar{x}(N|k_j)\|_P^2,$$

and

$$\tilde{J}_1^a(k_j) = \sum_{i=0}^{N-1} \mathrm{tr}\{(Q + \bar{K}^{\mathrm{T}} R\bar{K})[\bar{\Phi}^i E(0|k_j)(\bar{\Phi}^i)^{\mathrm{T}}] $$
$$+ \sum_{k=0}^{i-1} \bar{\Phi}^k FWF^{\mathrm{T}}(\bar{\Phi}^k)^{\mathrm{T}}\} + \mathrm{tr}\{P\bar{\Phi}^N E(0|k_j)(\bar{\Phi}^N)^{\mathrm{T}}\} + \sum_{i=0}^{N-1} \bar{\Phi}^i FWF^{\mathrm{T}}(\bar{\Phi}^i)^{\mathrm{T}}.$$

It can be readily shown that

$$J_{l^*(k_j)}^*(k_j) \le J_1^*(k_j) \le \bar{J}_1^a(k_j) + \tilde{J}_1^a(k_j) = \mathbb{E}\{\|x(k_j)\|_P^2\} + N\mathrm{tr}\{PFWF^{\mathrm{T}}\}$$
$$\le \bar{\lambda}(P)\mathbb{E}\{\|x(k_j)\|^2\} + N\mathrm{tr}\{PFWF^{\mathrm{T}}\}. \tag{2.36}$$

If $(\bar{x}_{0|k}, E_{0|k}) \in \Omega_F$, then from (2.34) and (2.36), we have

$$J_1^*(k_{j+1}) \le J_{l^*(k_j)}^*(k_j)\left(1 - \frac{1}{\alpha}\frac{\underline{\lambda}(Q)}{\bar{\lambda}(P)}\right) + \left(\frac{1}{\alpha}\frac{\underline{\lambda}(Q)}{\bar{\lambda}(P)}N + l^*(k_j)\right)\mathrm{tr}\{PFWF^{\mathrm{T}}\}$$
$$- \frac{1}{\alpha}\sum_{i=0}^{l^*(k_j)-1} \mathrm{tr}\{(l^*(k_j) - i - 1)QA^iFW(A^iF)^{\mathrm{T}}\}. \tag{2.37}$$

If $J_{l^*(k_j)}(k_j) \le \rho^{-1}\beta_{\max}\mathrm{tr}\{PFWF^{\mathrm{T}}\}$, from (2.29) and (2.35), we have

$$\mathbb{E}\{\|x(0|k_j)\|^2\} = \|\bar{x}(0|k_j)\|^2 + \mathrm{tr}(E(0|k_j)) \le \frac{\beta_{\max}}{\rho\underline{\lambda}(Q)}\mathrm{tr}(PFWF^{\mathrm{T}}). \tag{2.38}$$

Recalling (2.28), we have

$$\|\bar{x}(0|k_j)\|^2 < \bar{\sigma}^2, \mathrm{tr}(E(0|k_j)) < \underline{\lambda}(\bar{E}), \tag{2.39}$$

which implies that $(\bar{x}(0|k_j), E(0|k_j))$ is in the interior of the terminal set Ω_F. Recalling (2.37), we can guarantee if $J_{l^*(k_j)}(k_j) \le \rho^{-1}\beta_{\max}\mathrm{tr}\{PFWF^{\mathrm{T}}\}$, then $J_{l(k_{j+1})}(k_{j+1}) \le$

$J^*_{l^*(k_j)}(k_j) \leq \rho^{-1}\beta_{\max}\mathrm{tr}\{PFWF^{\mathrm{T}}\}$. Based on this, we can establish a positive invariant set D as

$$D = \{(\bar{x}_{k_j}, E_{k_j}) : J_{l(k_j)}(k_j) \leq \rho^{-1}\beta_{\max}\mathrm{tr}(PFWF^{\mathrm{T}})\}. \tag{2.40}$$

The following proof is similar as the one in [133]. For $(\bar{x}(0|k_j), E(0|k_j)) \in \Omega_F\backslash D$, it holds that

$$J_{l^*(k_j)}(k_j) > \rho^{-1}\beta_{\max}\mathrm{tr}(PFWF^{\mathrm{T}}). \tag{2.41}$$

Considering (2.34), (2.36), we can get

$$J_{l^*(k_{j+1})}(k_{j+1}) - J_{l^*(k_j)}(k_j) < 0. \tag{2.42}$$

Meanwhile, for all $(\bar{x}(0|k_j), E(0|k_j)) \in \Xi\backslash\Omega_F$, where Ξ is the initial feasible set, there exist constant $\bar{c}$ and set $(\bar{x}_\Omega, E_\Omega) \in \Omega_F\backslash D$ such that $-\underline{\lambda}(Q)\mathbb{E}\{\|x(k_j)\|^2\} \leq -\underline{\lambda}(Q)\mathbb{E}\{\|x_\Omega\|^2\} - \bar{c}$. This implies that for all $(\bar{x}(0|k_j), E(0|k_j)) \in \Xi\backslash\Omega_F$,

$$J_{l^*(k_{j+1})}(k_{j+1}) - J_{l^*(k_j)}(k_j) < -\bar{c}. \tag{2.43}$$

This implies that there exists $T_1 > 0$ such that $(\bar{x}(0|k_j+T_1), E(0|k_j+T_1)) \in \Omega_F$. From the invariance of the set D, $(\bar{x}(0|k_j+T_1), E(0|k_j+T_1)) \in D$ implies $(\bar{x}(0|k_j), E(0|k_j)) \in D$ for all $k_j \geq T_1$. If $(\bar{x}(0|k_j + T_1), E(0|k_j + T_1)) \in \Omega_F\backslash D$, we have

$$J_{l^*(k_{j+1})}(k_{j+1}) - J_{l^*(k_j)}(k_j) \leq -(1-\rho)\frac{\underline{\lambda}(Q)^2}{\bar{\lambda}(P)}\mathbb{E}\{\|x_{k_j+T_1}\|^2\}. \tag{2.44}$$

Then for all $\epsilon > 0$, there exists $T_2 \geq T_1$ such that,

$$J_{l^*(k_j)}(k_j) \leq \epsilon + \rho^{-1}\beta_{\max}\mathrm{tr}(PFWF^{\mathrm{T}}),$$

for all $k \geq T_2$. Considering (2.35), it can be shown that

$$d(\mathbb{E}\{\|x(k_j)\|^2_Q\}, [0, \rho^{-1}\beta_{\max}\mathrm{tr}(PFWF^{\mathrm{T}})]) \to 0$$

as $j \to +\infty$. $\qquad\square$

Remark 4. *In the proof, we need to assume that A is non-singular, and this is always true if we discretize the continuous-time linear system by a sample-and-hold method to get the system in (2.1).*

Remark 5. *From Theorem 2, it can be observed that the stability condition of the designed stochastic self-triggered MPC algorithm is to check the existence of a ρ that satisfies (2.28). β_{max} defined in (2.29) can be readily decided offline, and it has been shown that β_{max} can be determined by the inter-execution time interval l and the open-loop dynamics of the system. $\bar{\sigma}^2$ in the right hand side of (2.28) defines a ball in the state terminal region $\mathbb{X}_F$, and the maximum value of $\bar{\sigma}$ is restrained by $\mathbb{X}_F$. The terminal variance constraint $\bar{E}$ is related to $\bar{W}$, which is a design parameter defined in (2.20).*

2.5   Numerical examples

In this section, two examples are given to demonstrate the reduction of communication load by applying the stochastic self-triggered MPC control method. Simulations were accomplished in MATLAB on a Laptop with a 2.6 GHz Intel Core i7 CPU and 16 GB RAM. We used Yalmip [134] and SDPT3 as the QP solver to solve the optimization problem.

2.5.1   Comparison between the self-triggered and the periodically triggered stochastic MPC

Let the system be given as follows:

$$x(k+1) = \begin{bmatrix} 1.1 & 1 \\ 0 & 1.2 \end{bmatrix} x(k) + \begin{bmatrix} 0.5 \\ 1 \end{bmatrix} u(k) + \begin{bmatrix} 1 & 0 \\ 0 & 1 \end{bmatrix} w(k), k \in \mathbb{N}_{\geq 0}$$

in which $x(k)$ and $u(k)$ are subject to the state constraint $x(k) \in \mathbb{X} = [-20, 20] \times [-8, 8]$, input constraint $u(k) \in \mathbb{U} = [-8, 8]$, respectively. The disturbance vector $w(k)$ is assumed to be normally distributed with a zero mean, and a variance $W = \text{diag}(1/12, 1/12)$. The weighting matrices are selected to be $Q = \text{diag}(1, 1)$ and $R = 0.1$. $\bar{K} = \begin{bmatrix} -0.745 & -1.4270 \end{bmatrix}$ is chosen to be linear quadratic optimal with the weighting matrices Q and R. The state and input chance constraints (2.10) with $p = 0.3$ are considered in the simulation. We select $\epsilon = 0.3$ as the linearization factor of the chance constraints. The prediction horizon N is selected to be 10, and the maximum open-loop phase $L_{\max}$ is selected to be 5. To satisfy the stability condition (2.28), $\bar{W}$ is chosen as $\bar{W} = 10W$ for a larger terminal constraint $\bar{E}$ in (2.19).

Convergence of the state to the invariant set D**.** We first investigate the sampling frequency of the system with the proposed scheme. The initial state and initial variance are chosen as $x(0|k_j) = \begin{bmatrix} 0 & 0 \end{bmatrix}^{\mathrm{T}}$ and $E(0|k_j) = \underline{0}$. For $\alpha = 2$, a simulation of $T_{\mathrm{sim}} = 10^4$ steps yields the state trajectories as shown in Figure 2.2a. The light grey region is the terminal set $\bar{\mathbb{X}}_f$, and the dark grey region is the state constraint set $\mathbb{X}$. By selecting a $\rho \in (0, 1)$, the condition (2.28) can be readily verified. The red circle centred at the origin is the invariant set D with the smallest radius $\beta_{\mathrm{max}}\mathrm{tr}(PFWF^{\mathrm{T}})$. The zoomed-in part of the set D and the state trajectories are shown in Figure 2.2b, and it shows the invariance of the set D. For different values of α, the average inter-execution time and distribution of the inter-execution times are illustrated in Table 2.1. The average inter-execution time can be evaluated by $T_{\mathrm{sim}}/(\text{total sampling times})$. Note that, as stated in (2.29), since β_{max} is related to the value of α and W, to guarantee $D \in \bar{\mathbb{X}}_f$, the value of α cannot be selected too large.

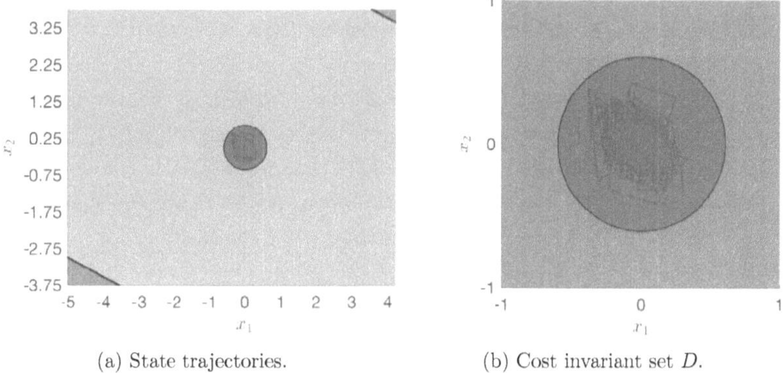

(a) State trajectories.

(b) Cost invariant set D.

Figure 2.2: State trajectories of the closed-loop system for the stochastic self-triggered MPC scheme with the initial point $[0, 0]^{\mathrm{T}}$.

Constraints violation in the transient response. To investigate the transient behaviour of the closed-loop system, we randomly select 300 initial conditions in the set $\mathbb{X}\backslash\bar{\mathbb{X}}_f$, and all initial conditions that result in an infeasible solution are excluded. For each initial condition, the closed-loop system is simulated with both self-triggered controller and periodically triggered controller for T_{sim} steps. The realizations of the disturbance are identical for these two controllers. As a result, the average inter-

Table 2.1: Distribution of average inter-execution time for the stochastic self-triggered MPC schemes with 100 different realizations of uncertainties.

α	Aver. inter-exe. time	Frequency of inter-exe. time l				
		1	2	3	4	5
1	1	100%	0	0	0	0
2	1.10	81.1%	18.1%	0.8%	0	0
3	1.25	76.2%	22.3%	1.4%	0.1%	0
4	1.35	68.6%	27.9%	3.1%	0.4%	0
6	1.43	63.9%	29.3%	6.46%	0.4%	0
12	1.55	56.1%	34.5%	7.5%	0.6%	1.3%
14	1.58	54.8%	35.3%	7.4%	2.1%	0.5%

execution time for the self-triggered stochastic MPC controller is 3.35, which amounts to a 70% reduction in the communication load compared to the periodically sampling scheme. For a randomly selected initial condition, a set of state trajectories of the closed-loop system for 100 realizations of the disturbance sequence are plotted in Figure 2.3. In Step 1, 10% of the state trajectories violates the state constraint, while in Step 2, 25% of the trajectory set violates the state constraint.

Comparison with the robust self-triggered MPC. In addition, the proposed stochastic self-triggered MPC scheme is compared with the tube-based robust self-triggered MPC scheme [132]. The performance index is defined as

$$J_{\text{perf}} = \frac{1}{T_{\text{sim}}} \sum_{k=0}^{T_{\text{sim}}-1} (\|x(k)\|_Q^2 + \|u(k)\|_R^2). \tag{2.45}$$

300 closed-loop simulations with a simulation length $T_{\text{sim}} = 50$ are conducted to evaluate the performances of two algorithms. In each simulation, the same realizations of uncertainty are used for both schemes. The average performance for the stochastic self-triggered MPC is 11.04, as compared with 12.33 for the robust self-triggered MPC. The average inter-execution time for robust self-triggered MPC is 2.55, while it is 2.34 for the proposed stochastic self-triggered MPC method. It can be concluded that the performance is improved without sacrificing much in communication reduction.

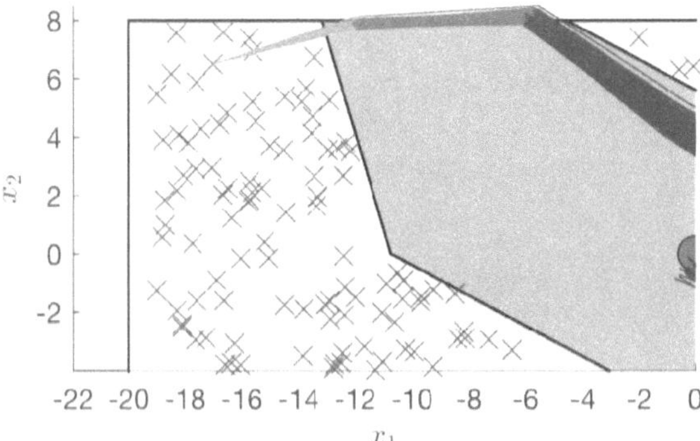

Figure 2.3: Blue crosses denote random initial conditions in the simulation. The solid lines denote the state trajectories of the closed-loop system for the stochastic self-triggered MPC scheme with the initial point $[-17, 6.5]^{\mathrm{T}}$. The grey region denotes the terminal constraints, and the red region is the cost invariant set.

2.5.2   2-D point-mass double-integrator plant

In the second example, we implement the designed stochastic self-triggered MPC algorithm to a two dimensional point-mass double-integrator system with uncertainty in positions. This model has already been considered in [59], and the control objective is to regulate the states of the system to the origin while reducing the communication load between the sensor and controller. The double-integrator system can be described as (2.1) in which

$$A = \begin{bmatrix} 1 & 0 & 1 & 0 \\ 0 & 1 & 0 & 1 \\ 0 & 0 & 1 & 0 \\ 0 & 0 & 0 & 1 \end{bmatrix}, B = \begin{bmatrix} \frac{1}{2} & 0 \\ 0 & \frac{1}{2} \\ 1 & 0 \\ 0 & 1 \end{bmatrix}, F = \begin{bmatrix} 0.1 & 0 \\ 0 & 0.1 \\ 0 & 0 \\ 0 & 0 \end{bmatrix}.$$

The positions of the system are constrained in a square with vertices $(-5, -5)$, $(5, -5)$, $(-5, 5)$, $(5, 5)$. The input constraint is given by $\|u(k)\|_\infty \le 1$. The disturbance $w(k)$

is an independent and identically distributed white noise with zero mean and variance matrix $W = I_2$. The weighting matrices are selected to be $Q = 10^{-4} \times \mathrm{diag}(1, 1, 1, 1)$ and $R = 1$. Comparing to R, Q is chosen significantly smaller since the optimal solutions will push the state to boundaries of chance constraints. $p = 0.2$ is selected for the state and input chance constraints (2.10) and the linearization factor is chosen to be $\epsilon = 0.2$. N and $L_{\max}$ are selected to be the same as the previous example. $\bar{K}$ is obtained as the optimal LQ gain, and $\bar{E}$ is calculated by using (2.20) with $\bar{W} = W$.

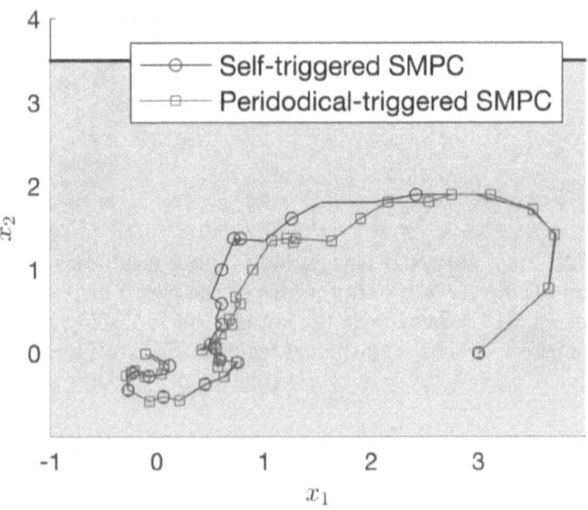

Figure 2.4: State trajectories under two implementation of stochastic MPC with the same initial point $[3, 0, 0, 0]^{\mathrm{T}}$. The blue circle and red square denote the sampling instants for self-triggered scheme and periodical-triggered scheme, respectively.

Figure 2.4 shows the state trajectories under periodical-triggered and self-triggered stochastic MPC for the same realization of uncertainty, respectively. For each scheme, 1000 closed-loop simulations with $\mathrm{T_{sim}} = 40$ steps have been performed. For both schemes, it can be observed that the states can be stabilized around the origin. The average inter-execution time for the self-triggered stochastic MPC is 2.22. Comparing to the periodical triggered scheme, the communication cost is reduced by 55%. The performance index is defined as (2.45), and it can be calculated that the performance index for the self-triggered scheme is 10% smaller than the periodical-triggered scheme. It can be observed the sampling rate for the stochastic self-triggered MPC is larger than the periodically sampling case without much loss in performance.

2.6 Conclusions

In this chapter, a stochastic self-triggered MPC scheme is proposed for linear constrained discrete-time systems. The proposed self-triggered sampling scheme effectively reduces the communication load between the sensor and the controller because of the implementation of the self-triggered sampling scheme. The recursive feasibility of the proposed control scheme and the stability conditions are developed. Simulation results have demonstrated the effectiveness of the algorithm.

Chapter 3

Stochastic Self-triggered MPC with Adaptive Prediction Horizon for Linear Systems subject to Chance Constraints

3.1 Introduction

Recently, stochastic model predictive control (SMPC) has received great attention for constrained control of systems subject to stochastic disturbances and chance constraints. Different from robust MPC (RMPC) approaches, in which worst-case realizations of disturbances are assessed explicitly, SMPC utilizes the stochastic characterization of disturbances to relax the inherent conservativeness of RMPC. Particularly, SMPC offers less conservative treatment of constraints by exploiting the probabilistic distribution information of disturbances to define chance constraints, which allows some degree of constraint violations. Based on chance constraints handling methods, SMPC algorithms in the literature can be broadly classified into two approaches: Analytic approximation approach [43, 69, 121], and randomized approach [61, 51]. In the first formulation, chance constraints are offline reformulated into deterministic form by exploiting the distributional information of disturbances, leading to linear constraints imposed on nominal prediction dynamics. This method results in a conservative approximation to the original chance-constrained optimization problem, but the decision variables in the online optimization problem are significantly reduced.

Closed-loop properties, such as the convergence of states, recursive feasibility, and chance constraints satisfaction, have been established for linear systems subject to either bounded [43] or unbounded [72] disturbances. One limitation of the analytic approximation approach is that the reformulation relies on specific disturbance distributions. On the other hand, the randomized approach depends on the scenario-based optimization technique [61] and the main feature of this approach is to utilize appropriate sampling of constraints to approximate the original stochastic optimal control problem. It offers flexible applicability to a broader class of SMPC problems since no probability distribution of uncertainty is assumed to be known as long as samples of uncertainties can be obtained. Generally speaking, this approach is computationally intensive, especially for large scale systems, and the recursive feasibility of the MPC algorithm is not well investigated, except for the recent work [49] which combines the scenario approach and robust constraint tightening technique. Interested reader please refer to [57, 135] for a detailed review of SMPC.

Modern networked control systems or cyber-physical systems are subject to ubiquitous physical constraints, model uncertainties, external disturbances and indispensable communication constraints, which can severely degrade the control performance and even destabilize the closed-loop system. The event-based MPC is an ideal approach for controlling this type of system due to advantages such as achieving optimal performance, handling state and control input constraints explicitly, and reducing the communicational load for networked systems. As a result, the event-based MPC [100, 136], including event-triggered MPC and self-triggered MPC, has been extensively investigated in recent years. In event-triggered MPC, a pre-designed triggering condition, based on the error between the real system state and predicted one, will be checked continuously to determine whether the MPC control update is triggered or not. Differently, in self-triggered MPC, the next triggering time instant is precomputed at the current sampling time by designing an appropriate self-triggering condition using the system dynamics and state predictions. This feature overcomes the drawback in event-triggered MPC, where the continuous checking of triggering conditions maybe not practical. The self-triggered MPC control method has been developed in recent years, and most of the proposed strategies [137, 114, 138] are for systems without uncertainties or disturbances. When model uncertainties or disturbances are taken into account, existing self-triggered algorithms focus mainly on robust constraint satisfaction. Inspired by the tube-based MPC method, a class of robust self-triggered MPC is proposed in [113], where the triggering condition is designed based on the bounds of the MPC value function. The same triggering condition is also adopted in [139], where a robust min-max self-triggered MPC method is proposed. Later in [140] and [141], robust self-triggered MPC strategies based on reachability analysis and relaxed dynamic programming method are reported, respectively.

As discussed above, the integration of SMPC and self-triggered MPC can ensure closed-loop chance constraints satisfaction and therefore reduce the inherent conservativeness of RMPC. Alternative stochastic self-triggered MPC schemes are available in the literature, such as [110, 111, 112]. In [110] and [111], the similar self-triggering condition inspired by [113] is adopted, while different chance constraints handling methods from [43] and [72] are utilized, respectively. In [110], the stability of the system is analyzed in the mean-square sense while input-state stability of the closed-loop system is proved in [111]. In [112], the self-triggering condition is designed based on the summation of the MPC value function bound and the performance measure at the last sampling time instant. One common feature of previously mentioned works on self-triggered SMPC is that the open-loop control paradigm is applied between triggering time instants, and resulting constraint tightenings are therefore more complex and time-varying compared to conventional SMPC methods. The triggering condition design all relies on the bounds of the MPC value function, where a periodical sampling is assumed after the open-loop phase. In addition, to evaluate the triggering condition at each sampling time instant, the solution to a set of quadratic programs with time-varying tightened constraints is required. Thus, another fundamental limitation is that all methods are essentially computationally expensive at each sampling time instants.

In this chapter, the design of a self-triggered SMPC algorithm is considered for linear systems subject to additive disturbances and chance constraints. The self-triggering mechanism proposed in [138] explicitly handles the communication effect in the cost function by adding a damping factor, and both the control inputs and next triggering time are optimized designed simultaneously. One unique feature in [138] is that the prediction horizon is designed as a variable associated with inter-execution time, and therefore the patterns of control sequence change adaptively. The self-triggering mechanism is then further extended to the distributed case in [142]. However, both [138] and [142] consider undisturbed systems, which omit the ubiquitous uncertainties in practice. Since disturbances are not considered in previous works [138], [142] under the adaptive triggering condition, some appropriate modifications are required: (i) a dynamic feedback gain selection procedure is designed to compensate for disturbance propagations in the open-loop operations, leading to a more complex error prediction formulation to construct the tightening parameters; (ii) a modified terminal constraint is imposed to the terminal state in order to guarantee the recursive feasibility of the scheme in spite of the dynamic gain selection; (iii) an improved self-triggering condition with more tunning parameters is proposed to evaluate the tradeoff between control performance and communication cost explicitly. In summary, the contribution of this chapter is in two-folds:

- The co-design of self-triggering mechanism and SMPC algorithm can effectively reduce both the communication and computation burden, while the sacrifice of control performance is guaranteed within some specific levels. The presented self-triggering mechanism offers greater flexibility in tunning the triggering behaviour in the presence of state chance constraints and external disturbances.

- The probability of state constraint violations is tight to the desired value under the self-triggered mechanism. Also, the recursive feasibility of the proposed method and closed-loop stability of the system have been analyzed in this work.

The structure of this chapter is organized in the following way. In Section 3.2, the problem formulation is given and a prototype SMPC controller is defined. Following that in Section 3.3, the modified cost function and constraint tightening methods are introduced. In Section 3.4, the self-triggered stochastic MPC problem is defined, and closed-loop properties of the system under the control scheme are analyzed. The numerical example is given in Section 3.5 and Section 3.6 concludes this chapter.

Notations: In the following, $\mathbb{N}$ denotes the set of integers and $\mathbb{R}$ denotes the set of real numbers. $\mathbb{N}_0$ denotes the set of natural number. For some integers $a \leq x \leq b$ is denoted as $x \in \mathbb{N}_{[a,b]}$. The matrix I and $\mathbf{0}$ denote identity matrix and zero matrix with some appropriate dimensions. For a vector $x(k) \in \mathbb{N}^{n_x}$, the $A-$ weighted norm is written as $\|x(k)\|_A^2 = x^{\mathrm{T}} A x$. $A \succ 0$, $A \succeq 0$ denote the matrix A is positive definite, semi-definite positive, respectively. $x(i|k)$ denotes the i-step ahead predicted state given the state $x(k)$. For a random variable x, $\Pr(x)$ and $\mathbb{E}(x)$ denote the probability and expected value of x.

3.2 Problem setup

In this section, we will introduce the problem formulation of the stochastic self-triggered MPC with adaptive prediction horizon, including the system dynamics, self-triggered mechanism, control objectives, and resulting prototype optimization problems.

3.2.1 System dynamics and chance constraints

Consider the following linear time-invariant system subject to stochastic additive disturbances:

$$x(k+1) = Ax(k) + Bu(k) + w(k), k \in \mathbb{N}_0, \tag{3.1}$$

in which $x(k) \in \mathbb{R}^{n_x}$ denotes the state, $u(k) \in \mathbb{R}^{n_u}$ denotes the control input and $w(k) \in \mathbb{R}^{n_w}$ denotes stochastic additive disturbance. The disturbance sequence $\{w(0), \ldots, w(k), \ldots\}$ is a realization of random process $W(k), k \in \mathbb{N}_0$ satisfying the following assumption.

Assumption 2. *The random process $W(k)$ is independent, identically distributed with zero mean, and is supported by $\mathbb{W}$, in which $\mathbb{W}$ is a bounded and convex set. The probability distribution of $W(k)$ is given by $F(\cdot)$.*

Moreover, the system is subject to chance constraints on predicted states in the form of

$$\Pr(g^{\mathrm{T}}x(i+1|k) \leq h) \geq p_x, i \in \mathbb{N}_0, \tag{3.2}$$

where p_x is the constraint violations probability level. The predicted state at time instant $k+i+1$ is modeled as:

$$x(i+1|k) = Ax(i|k) + Bu(i|k) + W(k+i), i \in \mathbb{N}_0,$$

in which $x(i|k)$ and $u(i|k)$ are predicted state and control input given $x(0|k) = x(k)$.

3.2.2 Self-triggered mechanism

In conventional SMPC, a widely used control parameterization of predicted inputs $u(i|k)$ is in the form of

$$u(i|k) = Kx(i|k) + c(i|k), i \in \mathbb{N}_0, \tag{3.3}$$

in which $c(i|k), i \in \mathbb{N}_{[0,N-1]}$ are optimization variables, and $c(i|k) = 0, i \in \mathbb{N}_{\geq N}$ with a given prediction horizon N. By solving the stochastic MPC problem, the control input is then applied to the system in a receding horizon fashion, as shown in Figure 3.1a. In contrast, in the self-triggered scheme with fixed prediction horizon N, control inputs are not generated and transmitted at each sampling time instant k. Define the triggering index as $j \in \mathbb{N}_0$, and the time sequence for control input updates as $\{k_0, k_1, \ldots, k_j, \ldots\}$ with $k_0 = 0$. As shown in Figure 3.1b, at sampling time instant k_j, the next sampling time instant is defined as $k_{j+1} = k_j + \tau_j$, where $\tau_j \in \mathbb{N}_{[1,N-1]}$ is defined as the inter-execution time determined by the self-triggered mechanism given the state $x(k_j)$. The main objectives of self-triggered mechanism design are to determine the next sampling time instant k_{j+1} and control actions during the period $k_{j+1} - k_j$, given the system information at time instant k_j.

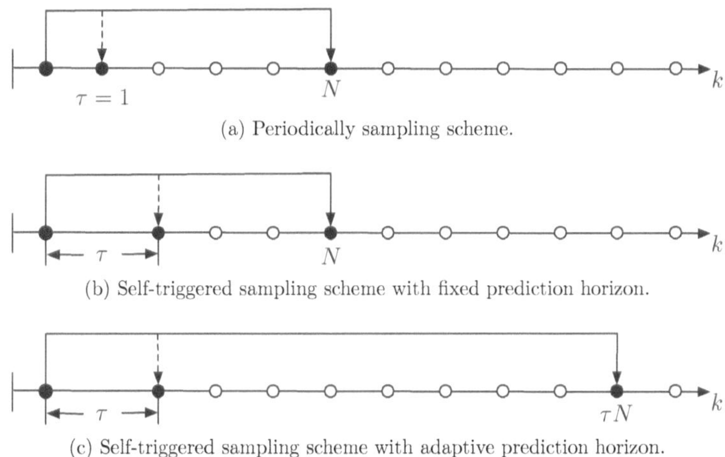

(a) Periodically sampling scheme.

(b) Self-triggered sampling scheme with fixed prediction horizon.

(c) Self-triggered sampling scheme with adaptive prediction horizon.

Figure 3.1: Comparions of periodically sampling scheme, self-triggered mechanisms with fixed prediction horizon and adaptive prediction horizon.

Inspired by [138], the prediction horizon is formulated as a variable that is related to the inter-execution time τ, as shown in Figure 3.1c. In this case, the cost function in the stochastic paradigm at k_j corresponding to some fixed inter-execution time instants $\tau \in \mathbb{N}_{[1,N]}$ can be defined as

$$
\begin{aligned}
J(x(k_j), \mathbf{c}^{[\tau]}(k_j), \tau) &= e^{-\gamma\tau}\mathbb{E}\left\{ \sum_{i=0}^{N\tau-1} \|x(i|k_j)\|_Q^2 + \|u(i|k_j)\|_R^2 + \|x(N\tau|k_j)\|_P^2 \right\} \\
&= e^{-\gamma\tau}\mathbb{E}\left\{ \sum_{l=0}^{N-1} L^{[\tau]}(x(l\tau|k_j), \mathbf{c}^{[\tau]}(l|k_j), \tau) + \|x(N\tau|k_j)\|_P^2 \right\},
\end{aligned}
\tag{3.4}
$$

where the stage function corresponding to τ is defined as $L(x(l\tau|k_j), \mathbf{c}^{[\tau]}(l|k_j), \tau) = \sum_{s=0}^{\tau-1} \|x(l\tau + s|k_j)\|_Q^2 + \|u(l\tau + s|k_j)\|_R^2$. The scalar $\gamma > 0$ is a tuning parameter and $e^{-\gamma\tau}$ is therein the damping factor characterizing the communication cost. $Q \succeq 0, R \succ 0$ and $P \succ 0$ are symmetric weighting matrices. For any τ, patterns of control perturbation sequence $\mathbf{c}^{[\tau]}(k_j)$ are defined as

$$
\mathbf{c}^{[\tau]}(k_j) =
$$
$$
\left[\underbrace{c(0|k_j)^{\mathrm{T}}, \ldots, c(0|k_j)^{\mathrm{T}}}_{\tau}, \ldots, \underbrace{c(l|k_j)^{\mathrm{T}}, \ldots, c(l|k_j)^{\mathrm{T}}}_{\tau}, \ldots, \underbrace{c(N-1|k_j)^{\mathrm{T}}, \ldots, c(N-1|k_j)^{\mathrm{T}}}_{\tau} \right]^{\mathrm{T}}.
\tag{3.5}
$$

When $\tau = 1$, the communication frequency is the highest for the reason that the least damping $e^{-\gamma}$ is given to the cost function (3.4). As τ increases, the damping factor on the cost increases, while on the other hand, the cost function value also increases, leading to a poorer control performance. The existence of a possible optimal τ represents the tradeoff between the communication cost and control performance.

Remark 6. *Note that the cost function defined in (3.4) is different from that in [138] because an expectation form of the cost function is adopted in this work. Different from the cost function in [110] where a penalty scalar is introduced in the open-loop phase, a damping factor $e^{-\gamma\tau}$ is introduced in the cost to take the communication cost into account explicitly. Also, the terminal cost $\|x(N\tau|k_j)\|_P^2$ is added to approximate the infinite horizon cost.*

At sampling time instant k_j, the prototype finite horizon optimal control problem

$\mathbb{P}_0^{[\tau]}(x(k_j), \mathbf{c}^{[\tau]}(k_j))$ given $x(k_j)$ is defined as

$$
\begin{aligned}
\min_{\{\mathbf{c}^{[\tau]}(k_j)\}} \quad & J(x(k_j), \mathbf{c}^{[\tau]}(k_j), \tau) \\
\text{s.t.} \quad & x(0|k_j) = x(k_j), \\
& x(i+1|k_j) = Ax(i|k_j) + Bu(i|k_j) + W(k_j + i), \quad i \in \mathbb{N}_0, \\
& \Pr(g^{\mathrm{T}}x(i+1|k_j) \le h) \ge p_x, \qquad\qquad\qquad i \in \mathbb{N}_0, .
\end{aligned}
\tag{3.6}
$$

To this end, the objective of this work is to design an appropriate self-triggered mechanism with modifications on the prototype optimization problem (3.6) to maximize the inter-execution time interval τ while guaranteeing chance constraints satisfaction (3.2) and closed-loop stability. Solving the optimization problem $\mathbb{P}_0^{[\tau]}(x(k_j), \mathbf{c}^{[\tau]}(k_j))$ online is generally computationally intractable because of the presence of probabilistic state constraints over infinite dimensions. To solve these challenging issues, some proper approximations and modifications are required to be carried out for $\mathbb{P}_0^{[\tau]}(x(k_j), \mathbf{c}^{[\tau]}(k_j))$, as shown in the next section.

3.3 Cost function and chance constraints handling

In this section, the chance-constrained optimization problem $\mathbb{P}_0^{[\tau]}(x(k_j), \mathbf{c}^{[\tau]}(k_j))$ is reformulated to a computationally tractable MPC problem by transforming the chance constraints into a deterministic form. Then, chance constraints handling methods are introduced, and recursive feasible stochastic tubes are constructed for predicted states $x(i|k_j)$. Finally, an improved terminal constraint is designed to deal with the infinite horizon constraints.

3.3.1 Receding horizon performance with terminal cost function

As discussed in aforementioned discussions, the prediction horizon in $\mathbb{P}_0^{[\tau]}(x(k_j), \mathbf{c}^{[\tau]}(k_j))$ changes adaptively with inter-execution time interval $\tau \in \mathbb{N}_{[1,\bar{\tau}]}$, where $\bar{\tau}$ is the maximum length of τ. Then, the control parametrization associated with τ is defined as

$$
u(l\tau + s|k_j) = K_s x(l\tau|k_j) + c(l|k_j), s \in \mathbb{N}_{[0,\tau-1]}, l \in \mathbb{N}_{[0,N-1]}.
\tag{3.7}
$$

Note that, if the inter-execution time τ is chosen as $\tau = 1$ (the system is sampled periodically), and a fixed feedback gain $K_s = K$ is chosen, then the control parameterization (3.7) is equivalent to (3.3). (3.7) emphasizes that the control parametrization relies on the predicted triggering time sequence $\{k_j, \ldots, k_j + l\tau, \ldots\}$. The predicted control actions during the inter-execution time interval depend on the sampling of state at k_j since the real state during the period is missing because of the self-triggering mechanism. Different from [110] where linear state feedback is introduced after the first τ steps, state feedback is introduced every τ steps because of the form of the control parametrization in (3.7).

Remark 7. *The control parametrization contains a set of state feedback gain $K_s, s \in \mathbb{N}_{[0,\tau-1]}$ to deal with the open-loop phase in the prediction. In previous works on self-triggered SMPC, such as [110, 111, 112], an important assumption is that state feedback is introduced to the prediction after the first τ steps. However, this is not the case in the proposed formulation, and a dynamic gain feedback K_s is therefore introduced. This dynamic parametrization of control input will reduce the conservativeness of the fixed state feedback gain scheme (3.3) in relative self-triggered mechanisms [138] and [110] significantly.*

At sampling time instant k_j, the control perturbation matrix $\mathbf{c}^{[\tau]}(k_j)$ for each inter-execution time $\tau \in \mathbb{N}_{[1,\bar{\tau}]}$ can be defined similarly as (3.5):

$$
\begin{aligned}
\mathbf{c}^{[\tau]}(k_j) &= [\mathbf{c}^{[\tau]}(0|k_j)^{\mathrm{T}}, \ldots, \mathbf{c}^{[\tau]}(N-1|k_j)^{\mathrm{T}}]^{\mathrm{T}}, \\
\mathbf{c}^{[\tau]}(l|k_j) &= [\underbrace{c(l|k_j)^{\mathrm{T}}, \ldots, c(l|k_j)^{\mathrm{T}}}_{\tau}]^{\mathrm{T}} \in \mathbb{R}^{n_u \times \tau}, l \in \mathbb{N}_{[0,N-1]}.
\end{aligned}
\tag{3.8}
$$

It should be noted that a spatial self-triggering strategy is adopted in this work, which means that the decision variable during triggering instants is constant. This strategy dramatically reduces the number of decision variables in the proposed work despite the prediction horizon increases as τ increases. Due to the linearity of system (3.1), the predicted state $x(l\tau + s|k_j)$ can be separated into predicted dynamics in nominal state $z(l\tau + s|k_j)$ and error state $e(l\tau + s|k_j)$ as

$$
x(l\tau + s|k_j) = z(l\tau + s|k_j) + e(l\tau + s|k_j).
\tag{3.9}
$$

With the control parametrization defined as (3.7), $z(i|k_j)$ and $e(i|k_j)$ can be expressed as, for $\tau \in \mathbb{N}_{[1,\bar\tau]}$ and $l \in \mathbb{N}_0$, $s \in \mathbb{N}_{[1,\tau]}$,

$$z(l\tau + s|k_j) = \Phi_{s-1} z(l\tau|k_j) + H_B^{[s]} \mathbf{c}^{[\tau]}(l|k_j), \tag{3.10a}$$

$$e(l\tau + s|k_j) = \Phi_{s-1} e(l\tau|k_j) + \sum_{n=0}^{s-1} A^{s-1-n} W(k_j + l\tau + n), \tag{3.10b}$$

where $H_B^{[s]} = \underbrace{\begin{bmatrix} A^{s-1}B & \cdots & B & 0 \end{bmatrix}}_{\tau}$ and $\mathbf{c}^{[\tau]}(l|k_j)$ are defined in (3.8). The matrix Φ_s is defined by iteration:

$$\Phi_s = A\Phi_{s-1} + BK_s, \quad \Phi_0 = A + BK_0.$$

From (3.10a) and (3.10b), the predicted system dynamics which is related to the system state with τ is given by, for $l \in \mathbb{N}_0$,

$$z((l+1)\tau|k_j) = \Phi_\tau z(l\tau|k_j) + H_B^{[\tau]} \mathbf{c}^{[\tau]}(l|k_j), \tag{3.11a}$$

$$e((l+1)\tau|k_j) = \Phi_\tau e(l\tau|k_j) + \sum_{n=0}^{\tau-1} A^{\tau-1-n} W(k_j + l\tau + n). \tag{3.11b}$$

where the matrix $H_B^{[\tau]}$ is defined as $H_B^{[\tau]} = \underbrace{\begin{bmatrix} A^{\tau-1}B & \cdots & B \end{bmatrix}}_{\tau}$.

Assumption 3. *The pair (A, B) is controllable and no eigenvalue of the matrix A is $\lambda(A) = 1$ and $\lambda(A^i) = 1$ for $i \in \mathbb{N}_{[1,\bar\tau]}$.*

Remark 8. *The design objective of the feedback gain K_s is to guarantee all $\Phi_s, s \in \mathbb{N}_{[1,\tau]}$ are Schur stable. One method is to solve the semi-definite programming problem iteratively to find feasible $K_s, s \in \mathbb{N}_{[1,\tau]}$ satisfying $\Phi_s^{\mathrm{T}} P_s + P_s \Phi_s \leq -I$. Gain selection methods with theoretical guarantees are subject to a future research topic.*

Proposition 1. *(Reformulation of stage function) For $\tau \in \mathbb{N}_{[1,\bar\tau]}$ and $l \in \mathbb{N}_{[0,N-1]}$, the expectation of stage function in (3.4) can be expressed as*

$$\begin{aligned}
&\mathbb{E}\{L(x(l\tau|k_j), \mathbf{c}^{[\tau]}(l|k_j), \tau)\} \\
&= \|z(l\tau|k_j)\|_{Q^{[\tau-1]}}^2 + 2z(l\tau|k_j)^{\mathrm{T}} N^{[\tau-1]} c(l|k_j) + \|c(l|k_j)\|_{R^{[\tau-1]}}^2 \\
&\quad + \mathbb{E}\left\{ \|e(l\tau|k_j)\|_{Q^{[\tau-1]}}^2 \right\} + \sum_{s=0}^{\tau-1} \mathrm{tr}\left(\sum_{n=0}^{s-1} A^n \Sigma^w (A^n)^{\mathrm{T}} Q \right),
\end{aligned} \tag{3.12}$$

where the matrices $Q^{[\tau-1]}$, $R^{[\tau-1]}$ and $N^{[\tau-1]}$ can be defined by the following iteration: For $s \in \mathbb{N}_{[0,\tau-1]}$,

$$
\begin{aligned}
Q^{[s]} &= Q^{[s-1]} + \Phi_s^{\mathrm{T}} Q \Phi_s + K_s^{\mathrm{T}} R K_s, \\
R^{[s]} &= R^{[s-1]} + (B^{[s-1]})^T Q B^{[s-1]} + R, \\
N^{[s]} &= N^{[s-1]} + \Phi_s^T Q B^{[s-1]} + K_s^{\mathrm{T}} R,
\end{aligned}
\tag{3.13}
$$

with $Q^{[0]} = Q + K_0^{\mathrm{T}} R K_0$, $R^{[0]} = R$, $N^{[0]} = K_0^T R$, and $B^{[s-1]} = \sum_{n=0}^{s} A^{s-1-n} B$.

By using the Proposition 1, the cost function in (3.4) can be expressed as

$$
\begin{aligned}
J&(x(k_j), \mathbf{c}^{[\tau]}(k_j), \tau) \\
=& e^{-\gamma \tau} \left(\sum_{l=0}^{N-1} \|z(l\tau|k_j)\|_{Q^{[\tau-1]}}^2 + 2z(l\tau|k_j)^{\mathrm{T}} N^{[\tau-1]} c(l|k_j) + \|c(l|k_j)\|_{R^{[\tau-1]}}^2 \right. \\
& + \sum_{l=0}^{N-1} \mathbb{E} \left\{ \|e(l\tau|k_j)\|_{Q^{[\tau-1]}}^2 \right\} + N \sum_{s=0}^{\tau-1} \mathrm{tr} \left(\sum_{n=0}^{s-1} A^n \Sigma^w (A^n)^{\mathrm{T}} Q \right) + \mathbb{E} \left\{ \|x(N\tau|k_j)\|_P^2 \right\} \bigg) \\
=& e^{-\gamma \tau} \left(\sum_{l=0}^{N-1} \|z(l\tau|k_j)\|_{Q^{[\tau-1]}}^2 + 2z(l\tau|k_j)^{\mathrm{T}} N^{[\tau-1]} c(l|k_j) + \|c(l|k_j)\|_{R^{[\tau-1]}}^2 + \|z_{N\tau|k_j}\|_P^2 \right) \\
& + W_{const}^{[\tau]},
\end{aligned}
\tag{3.14}
$$

where $W_{const}^{[\tau]}$ is a constant that doesn't depend on $x(k_j)$ and $\mathbf{c}^{[\tau]}(k_j)$ and can therefore be removed from the optimization problem, as shown in the Lemma 1 of [112]. Hence the modified cost function without constants is given by

$$
\begin{aligned}
\tilde{J}(x(k_j), \mathbf{c}^{[\tau]}(k_j), \tau) =& e^{-\gamma \tau} \left(\sum_{l=0}^{N-1} \|z(l\tau|k_j)\|_{Q^{[\tau-1]}}^2 + 2z(l\tau|k_j)^{\mathrm{T}} N^{[\tau-1]} c(l|k_j) \right. \\
& \left. + \|c(l|k_j)\|_{R^{[\tau-1]}}^2 + \|z_{N\tau|k_j}\|_P^2 \right).
\end{aligned}
\tag{3.15}
$$

Similar result can also be verified in the following statements on the construction of tightened constraints since the resulting tightened constraints are independent to $W_{const}^{[\tau]}$. The selection of the terminal cost weighting matrix P will be introduced in Section 3.3.3.

3.3.2 Chance constraints handling

In this subsection, we will convert the chance constraints into the deterministic form to render the prototype optimization problems $\mathbb{P}_0^{[\tau]}(x(k_j), \mathbf{c}^{[\tau]}(k_j))$ numerically tractable for every $\tau \in \mathbb{N}_{[1,\bar{\tau}]}$. Suppose the control parametrization (3.7) is implemented, the chance constraints (3.2) have an equivalent form as shown in the following lemma.

Lemma 2. *(Probabilistic constraint satisfaction) For $\tau \in \mathbb{N}_{[1,\bar{\tau}]}$, at sampling time instant k_j, the chance constraints $\Pr(g^{\mathrm{T}}x(l\tau + s|k_j) \leq h) \geq p_x$ are satisfied if and only if $\mathbf{c}^{[\tau]}(k_j)$ satisfies*

$$
\begin{aligned}
g^{\mathrm{T}}\Phi_{s-1}\Phi_\tau^l z(0|k_j) + g^{\mathrm{T}}\left(\Phi_{s-1}H_{\Phi_\tau}(l)\mathbf{c}^{[\tau]}(k_j) + H_B^{[s]}\mathbf{c}^{[\tau]}(l|k_j)\right) &\leq h \\
- \nu_{l\tau+i}^{[\tau]}, s \in \mathbb{N}_{[1,\tau]}, l \in \mathbb{N}_0,
\end{aligned}
\tag{3.16}
$$

where the matrix $H_{\Phi_\tau}(l)$ is defined as $H_{\Phi_\tau}(l) = \begin{bmatrix} \Phi_\tau^{l-1}H_B^{[\tau]} & \cdots & \Phi_\tau H_B^{[\tau]} & 0 & \cdots & 0 \end{bmatrix}$, and $\nu_{l\tau+s}^{[\tau]}$ is defined as the minimum value such that

$$
\begin{aligned}
\Pr\left\{g^{\mathrm{T}}\left[\Phi_{s-1}\left(\sum_{m=0}^{l-1}\Phi_\tau^m \sum_{n=0}^{\tau-1} A^{\tau-1-n}W(k_j + (l-1-m)\tau + n)\right)\right.\right. \\
\left.\left. + \sum_{n=0}^{s-1} A^{s-1-n}W(k_j + l\tau + n)\right] \leq \nu_{l\tau+s}^{[\tau]}\right\} = p_x.
\end{aligned}
\tag{3.17}
$$

Proof. By linearility of the system (3.1) and equation (3.10), the predicted state can be reformulated as for $s \in \mathbb{N}_{[1,\tau]}, l \in \mathbb{N}_0$,

$$
x(l\tau + s|k_j) = z(l\tau + s|k_j) + e(l\tau + s|k_j)
$$

$$
=\Phi_{s-1}z(l\tau|k_j) + H_B^{[s]}\mathbf{c}^{[\tau]}(l|k_j) + \Phi_{s-1}e(l\tau|k_j) + \sum_{n=0}^{s-1} A^{s-1-n}W(k_j + l\tau + n),
$$

$$
=\Phi_{s-1}\left(\Phi_\tau^l z(0|k_j) + \Phi_\tau^l e(0|k_j) + H_{\Phi_\tau}(l)\mathbf{c}^{[\tau]}(k_j)\right) + H_B^{[s]}\mathbf{c}^{[\tau]}(l|k_j)
$$

$$
+ \Phi_{s-1}\left(\sum_{m=0}^{l-1}\Phi_\tau^m \sum_{n=0}^{\tau-1} A^{\tau-1-n}W(k_j + (l-1-m)\tau + n)\right)
$$

$$
+ \sum_{n=0}^{s-1} A^{s-1-n}W(k_j + l\tau + n),
$$

where $z(l\tau|k_j) = \Phi_\tau z((l-1)\tau|k_j) + H_B^{[\tau]}\mathbf{c}_{l-1|k_j}^{[\tau]}$, and $e(l\tau|k_j) = \Phi_\tau e((l-1)\tau|k_j) +$

$\sum\limits_{n=0}^{\tau-1} A^{\tau-1-n} W(k_j + (l-1)\tau + n)$, are given by (3.11) for $l \in \mathbb{N}_{\geq 1}$. Considering the assumption on the initial error state $e(0|k_j) = 0$, it follows directly by (3.17) that (3.2) has an equivalent form as (3.16). $\qquad\qquad\square$

Remark 9. *The computation of $\nu_{l\tau+i}^{[\tau]}$ involves univariate convolutions which can be performed offline with arbitrarily small error, as suggested in [43]. Meanwhile, sampling-based approximation to $\nu_{l\tau+i}^{[\tau]}$ has also been proposed, as shown in [121].*

At sampling time instant k_j, the existence of $\mathbf{c}^{[\tau]}(k_j)$ implies only that future constraints will be satisfied with some given probability p_x. Since worst case realizations of uncertainties are not taken into account, the usual recursive feasibility cannot work in this case. As suggested in [43] and [110], at each sampling instant k_j, our concern is whether or not a feasible solution exists at the next sampling instant $k_{j+1} = k_j + \tau_j$. This depends not only on $x(k_{j+1})$, which in turn relies on $W(k_j), W(k_j + 1), \ldots, W(k_{j+1} - 1)$, but also on the assumption on the next inter-execution time interval τ_{j+1}. In [110] and [111], the predicted next sampling time interval τ_{j+1} is assumed to be 1, which indicates the system is sampled periodically after the first open-loop phase, while it is not the case in the proposed work. Similarly, worst case realizations for $W(k_j), W(k_j + 1), \ldots, W(k_{j+1} - 1)$ are considered to account for the $x(k_{j+1})$, and the following lemma provides conditions for the recursive feasibility of the algorithm to be asserted.

Theorem 3. *(Recursively feasible probabilistic tubes) At sampling time instant $k_j, j \in \mathbb{N}_0$, for any inter-execution time interval $\tau \in \mathbb{N}_{[1,\bar{\tau}]}$, consider the closed-loop dynamics*

$$
\begin{aligned}
x(k_j + i + 1) &= Ax(k_j + i) + Bu(k_j + i) + w(k_j + i), i \in \mathbb{N}_{[0,\tau]} \\
u(k_j + i) &= K_i x(k_j) + c(0|k_j).
\end{aligned}
\tag{3.18}
$$

Suppose there exists a control sequence $\mathbf{c}^{[\tau]}(k_j)$ satisfying

$$
\begin{aligned}
g^{\mathrm{T}} \Phi_{s-1} \Phi_\tau^l z(0|k_j) + g^{\mathrm{T}} \left(\Phi_{s-1} H_{\Phi_\tau}(l) \mathbf{c}^{[\tau]}(k_j) + H_B^{[s]} \mathbf{c}^{[\tau]}(l|k_j) \right) &\leq h \\
- \beta_{l\tau+s}^{[\tau]}, s \in \mathbb{N}_{[1,\tau]}, l \in \mathbb{N}_0,
\end{aligned}
\tag{3.19}
$$

where $\beta_{l\tau+i}^{[\tau]}$ is defined as the maximum element of the $(l\tau + i)$th column of the matrix

$\Gamma^{[\tau]}$ *in* (3.20).

$\Gamma^{[\tau]} =$

$$\begin{bmatrix} \nu_1^{[\tau]} & \cdots & \nu_\tau^{[\tau]} & \nu_{\tau+1}^{[\tau]} & \cdots & \nu_{2\tau}^{[\tau]} & \nu_{2\tau+1}^{[\tau]} & \cdots & \nu_{3\tau}^{[\tau]} & \cdots \\ 0 & \cdots & 0 & \nu_1^{[\tau]} + d_{\tau+1}^{[\tau]} & \cdots & \nu_\tau^{[\tau]} + d_{2\tau}^{[\tau]} & \nu_{\tau+1}^{[\tau]} + d_{2\tau+1}^{[\tau]} & \cdots & \nu_{2\tau}^{[\tau]} + d_{3\tau}^{[\tau]} & \cdots \\ \vdots & \cdots & \vdots & 0 & \cdots & 0 & \nu_1^{[\tau]} + d_{\tau+1}^{[\tau]} + d_{2\tau+1}^{[\tau]} & \cdots & \nu_\tau^{[\tau]} + d_{2\tau}^{[\tau]} + d_{3\tau}^{[\tau]} & \cdots \\ \vdots & \vdots & \vdots & \vdots & \vdots & \vdots & \vdots & \vdots & \vdots & \ddots \end{bmatrix},$$

$$(3.20)$$

with $d_{l\tau+i}^{[\tau]} = g^{\mathrm{T}} \left[\Phi_{s-1} \left(\sum_{q=0}^{l-1} \Phi_\tau^q \max_{w \in \mathcal{W}} \sum_{n=0}^{\tau-1} A^{\tau-1-n} w \right) \right], s \in \mathbb{N}_{[1,\tau]}, l \in \mathbb{N}_{\geq 1}$. *Then for*
the closed-loop system (3.18), *at the next sampling time instant* $k_{j+1} = k_j + \tau$, *there*
exists at least one feasible solution $\mathbf{c}^{[\tau]}(k_{j+1})$ *satisfying* (3.19). *Also, if future sampling*
time instants are assumed to be k_{j+m}, *where* $k_{j+m+1} = k_{j+m} + \tau$ *and* $m \in \mathbb{N}_0$, *the*
chance constraints (3.2) *are satisfied for all* $k \in \mathbb{N}_0$.

Proof. The definition of the first row in matrix $\Gamma^{[\tau]}$ makes (3.19) equivalent to (3.16).
At the next sampling time instant $k_{j+1} = k_j + \tau$, the feasibility of (3.19) is considered
by assuming $\tau_{j+1} = \tau$. At time k_{j+1}, define a candidate solution as $\tilde{\mathbf{c}}^{[\tau]}(k_{j+1}) = \left[\mathbf{c}^{[\tau]}(1|k_j)^{\mathrm{T}} \quad \cdots \quad \mathbf{c}^{[\tau]}(N-1|k_j)^{\mathrm{T}} \quad \mathbf{0} \right]^{\mathrm{T}}$, and it holds that $c(l\tau + s|k_{j+1}) = c((l+1)\tau + s|k_j)$ for $l \in \mathbb{N}_{[0,N-1]}, s \in \mathbb{N}_{[1,\tau]}$. The resulting predicted state at k_{j+1} for $l \in \mathbb{N}_0, s \in \mathbb{N}_{[1,\tau]}$, is given by

$$x(l\tau + s|k_{j+1}) = z(l\tau + s|k_{j+1}) + e(l\tau + s|k_{j+1})$$
$$= \Phi_{s-1} \left(\Phi_\tau^l z(0|k_{j+1}) + \Phi_\tau^l e(0|k_{j+1}) + H_{\Phi_\tau}(l)\tilde{\mathbf{c}}^{[\tau]}(k_{j+1}) \right) + H_B^{[s]}\tilde{\mathbf{c}}^{[\tau]}(l|k_{j+1})$$
$$+ \Phi_{s-1} \left(\sum_{m=0}^{l-1} \Phi_\tau^m \sum_{n=0}^{\tau-1} A^{\tau-1-n} W(k_{j+1} + (l-1-m)\tau + n) \right)$$
$$+ \sum_{n=0}^{s-1} A^{s-1-n} W(k_{j+1} + l\tau + n).$$

Note that from (3.11), $z(0|k_{j+1})$ and $e(0|k_{j+1})$ are given by $z(0|k_{j+1}) = \Phi_\tau z(0|k_j) + H_B^{[\tau]}\mathbf{c}^{[\tau]}(0|k_j)$, and $e(0|k_{j+1}) = \max_{w \in \mathcal{W}} \sum_{n=0}^{\tau-1} A^{\tau-1-n} w_{k_j+n}$, where $e(0|k_j) = 0$ and the worst case realizations of w in the period $[k_j, k_{j+1}]$ are considered. Hence, the candidate solution $\tilde{\mathbf{c}}(k_{j+1})$ is a feasible solution at time k_{j+1} if the following condition holds for $l \in \mathbb{N}_{\geq 1}, s \in \mathbb{N}_{[1,\tau]}$,

$$
\begin{aligned}
g^{\mathrm{T}} &\Phi_{s-1}\Phi_\tau^l z(0|k_j) + g^{\mathrm{T}}\left(\Phi_{s-1}H_{\Phi_\tau}(l)\mathbf{c}^{[\tau]}(k_j) + H_B^{[s]}\mathbf{c}^{[\tau]}(l|k_j)\right) \leq h \\
&- g^{\mathrm{T}}\left[\Phi_{s-1}\left(\Phi_\tau^l \max_{w \in \mathcal{W}} \sum_{n=0}^{\tau-1} A^{\tau-1-n}w\right)\right] \\
&- g^{\mathrm{T}}\left[\Phi_{s-1}\left(\sum_{m=0}^{l-1}\Phi_\tau^m \sum_{n=0}^{\tau-1} A^{\tau-1-n}W(k_{j+1}+(l-1-m)\tau+n)\right)\right. \\
&\left. \qquad + \sum_{n=0}^{s-1} A^{s-1-n}W(k_{j+1}+l\tau+n)\right],
\end{aligned}
\tag{3.21}
$$

which is the second row of the matrix $\Gamma^{[\tau]}$ and the subscripts in w are omitted for notation simplicity. Similarly, the feasibility of (3.19) at the sampling time instant $k_{j+p} = k_j + p\tau, p \in \mathbb{N}_0$ can be guaranteed if it holds that for $l \in \mathbb{N}_{\geq p}, s \in \mathbb{N}_{[1,\tau]}$,

$$
\begin{aligned}
g^{\mathrm{T}} &\Phi_{s-1}\Phi_\tau^l z(0|k_j) + g^{\mathrm{T}}\left(\Phi_{s-1}H_{\Phi_\tau}(l)\mathbf{c}^{[\tau]}(k_j) + H_B^{[s]}\mathbf{c}^{[\tau]}(l|k_j)\right) \leq h \\
&- g^{\mathrm{T}}\left[\Phi_{s-1}\left(\sum_{q=0}^{l-1}\Phi_\tau^q \max_{w \in \mathcal{W}} \sum_{n=0}^{\tau-1} A^{\tau-1-n}w\right)\right] \\
&- g^{\mathrm{T}}\left[\Phi_{s-1}\left(\sum_{m=0}^{l-1}\Phi_\tau^m \sum_{n=0}^{\tau-1} A^{\tau-1-n}W(k_{j+1}+(l-1-m)\tau+n)\right)\right. \\
&\left. \qquad + \sum_{n=0}^{s-1} A^{s-1-n}W(k_{j+p}+l\tau+n)\right],
\end{aligned}
\tag{3.22}
$$

which is the pth row of the matrix $\Gamma^{[\tau]}$. As a result, the feasibility at sampling instant $k_{j+m} = k_j + p\tau, p \in \mathbb{N}_0$, can be ensured by taking intersection of the above equations, and elements in those equations define matrix $\Gamma^{[\tau]}$ for $\tau \in \mathbb{N}_{[1,\bar{\tau}]}$. $\qquad\square$

Remark 10. *Theorem 3 implies that for some fixed $\tau_j \in \mathbb{N}_{[1,\bar{\tau}]}$, the optimization problem is always feasible for the following sampling time instant. By setting $\tau = 1$, the results are reduced to Theorem 3 in [43]. The key difference between the proposed Theorem 3 here and Theorem 3.1 in [110] is that $\tau_{k_j+m} = \tau_{k_j}, m \in \mathbb{N}_{\geq 1}$ in our work while $\tau_{k_j+m} = 1, m \in \mathbb{N}_{\geq 1}$ in [110].*

3.3.3 Terminal cost and terminal constraints

When $l \geq N$, the predicted control input is $u(l\tau + s|k_j) = K_s x(l\tau|k_j), s \in \mathbb{N}_{[0,\tau-1]}$, and the resulting predicted state dynamics evolves for $s \in \mathbb{N}_{[0,\tau-1]}, l \in \mathbb{N}_{\geq N}$ according to

$$x(l\tau + s + 1|k_j) = \Phi_s x(l\tau|k_j) + \sum_{n=0}^{s-1} A^{s-1-n} W(k_j + l\tau + n).$$

The selection of the terminal weighting matrix P follows the Section V in [138], where the similar result is extended to the stochastic setting.

Proposition 2. *(Terminal cost parameter design) If the terminal cost in (3.4) is designed as $\|x(N\tau|k_j)\|_P^2$, where $P \succ 0$ satisfies*

$$\Phi_{\tau-1}^{\mathrm{T}} P \Phi_{\tau-1} - P + Q^{[\tau-1]} \preceq 0, \forall \tau \in \mathbb{N}_{[1,\bar{\tau}]}, \tag{3.23}$$

then it holds that

$$\mathbb{E}\{\|x((N+1)\tau|k_j)\|_P^2 - \|x(N\tau|k_j)\|_p^2\} \leq -e^{\gamma\tau} L(x(N\tau|k_j), 0, \tau). \tag{3.24}$$

Remark 11. *The existence of a matrix P in (3.23) can be ensured by Lemma 3 in [138]. Since matrices Φ_τ are Hurwitz by construction for every $\tau \in \mathbb{N}_{[1,\bar{\tau}]}$, there exists a unique matrix $P^{[\tau]}$ satisfying $\Phi_\tau^{\mathrm{T}} P^{[\tau]} \Phi_\tau - P^{[\tau]} + \bar{Q}^{[\tau]} \prec 0$. Therefore the matrix P can be selected as $P = \sum_{\tau=1}^{\bar{\tau}} P^{[\tau]}$ following the similar result in [138]. The selection of P ensures that for all $\tau \in \mathbb{N}_{[1,\bar{\tau}]}$, the Proposition 2 is true, which is an extension of the Assumption 2 in [138] and plays a vital role in guaranteeing the stability of closed-loop system as shown in the Section 3.4.*

By using Theorem 3, the chance constraints (3.2) in the infinite horizon optimization problem $\mathbb{P}_0^{[\tau]}(\mathbf{c}_{k_j})$ can be transformed into the deterministic form. However, to render the problem numerically tractable, a terminal constraint has to be used to ensure the constraint (3.19) is satisfied over an infinite prediction horizon. In the following, for every $s \in \mathbb{N}_{[1,\tau]}$, the bounds on the sequence $\beta_{l\tau+s}^{[\tau]}$ are given first.

Lemma 3. *(Bounds on $\beta_{l\tau+s}^{[\tau]}$) For $\tau \in \mathbb{N}_{[1,\bar{\tau}]}$, and $s \in \mathbb{N}_{[1,\tau]}$, the sequence*

$$\beta_s^{[\tau]}, \beta_{\tau+s}^{[\tau]}, \ldots, \beta_{l\tau+s}^{[\tau]}, \ldots$$

is monotonically nondecreasing and upper bounded by

$$\bar{\beta}_s^{[\tau]} := \nu_s^{[\tau]} + \sum_{n=1}^{\lambda-1} d_{n\tau+s}^{[\tau]} + \frac{\rho_s^v}{1-\rho_s}\|g\|_{\Theta_s^{[\tau]}}, \tag{3.25}$$

where $\lambda \in \mathbb{N}_{>0}$, $\|g\|_{\Theta_s^{[\tau]}} = \sqrt{g^T \Theta_s^{[\tau]} g}$ *and* $\rho_s^{[\tau]}, \Theta_s^{[\tau]}$ *satisfy*

$$\max_{w \in \mathcal{W}} \|\sum_{n=0}^{\tau-1} A^n w\|_{(\Theta_s^{[\tau]})^{-1}} \leq 1, \tag{3.26a}$$

$$\Phi_{s-1}\Phi_\tau^l \Theta_s^{[\tau]}(\Phi_{s-1}\Phi_\tau^l)^{\mathrm{T}} \leq \rho_s^2 \Theta_s^{[\tau]}, \rho_s \in (0,1). \tag{3.26b}$$

Proof. From the definition of $\Gamma^{[\tau]}$ matrix, it follows that $\beta_{l\tau+s}^{[\tau]} \leq \beta_{(l+1)\tau+s}^{[\tau]}$ for every $s \in \mathbb{N}_{[1,\tau]}$, and we have

$$\lim_{l \to \infty} \beta_{l\tau+s}^{[\tau]} = \bar{\beta}_s^{[\tau]} \leq \nu_s^{[\tau]} + \sum_{l=1}^{\infty} d_{l\tau+s}^{[\tau]}. \tag{3.27}$$

It should be noted that both Φ_{s-1} and Φ_τ are strictly stable due to the control parametrization in (3.7), so conditions in (3.26) are feasible. From (3.26a), it holds that

$$\begin{aligned}
d_{l\tau+s}^{[\tau]} &= \max_{w \in \mathcal{W}} g^{\mathrm{T}} \Phi_{s-1}\Phi_\tau^l (\sum_{n=0}^{\tau-1} A^n w) \\
&\leq \max_{\|\lambda\|_{(\Theta_s^{[\tau]})^{-1}}} g^{\mathrm{T}} \Phi_{s-1}\Phi_\tau^l \lambda \\
&\leq \|(\Phi_{s-1}\Phi_\tau^l)^{\mathrm{T}} g\|_{\Theta_s^{[\tau]}}.
\end{aligned}$$

Therefore (3.26b) implies that $d_{l\tau+s}^{[\tau]} \leq \|(\Phi_{s-1}\Phi_\tau^{l-1})^{\mathrm{T}}g\|_{\Theta_s^{[\tau]}}$, which by recursion leads to (3.25) since $0 < \rho_s^{[\tau]} < 1$. $\qquad \square$

Remark 12. *To guarantee the existence of the recursive feasible tube (3.19), it is desirable to assume* $h \geq \bar{\beta}_s^{[\tau]}$.

From the predicted nominal dynamics in (3.10a), the terminal nominal dynamics can be written as

$$z((N+l)\tau+s|k_j) = \Phi_{s-1}\Phi_\tau^l z(N\tau|k_j), l \in \mathbb{N}_0, s \in \mathbb{N}_{[1,\tau]}, \tag{3.28}$$

since $c_{N+l|k_j} = 0$ for all $l \in \mathbb{N}_0$. The terminal constraint is therefore defined as

$$\mathcal{Z}_f^{[\tau]} := \left\{ z \in \mathbb{R}^{n_x} \;\middle|\; g^{\mathrm{T}} \Phi_{s-1} \Phi_\tau^l z \leq h - \beta_{N\tau+l\tau+s}^{[\tau]}, l \in \mathbb{N}_0, s \in \mathbb{N}_{[1,\tau]} \right\}. \tag{3.29}$$

From Lemma 3, the value of $\beta_{N\tau+l\tau+i}^{[\tau]}$ is upper bounded, and an inner approximation of (3.29) can be formulated as:

$$\hat{\mathcal{Z}}_f^{[\tau]} := \left\{ z \in \mathbb{R}^{n_x} \;\middle|\; \begin{array}{ll} g^{\mathrm{T}} \Phi_{s-1} \Phi_\tau^l z \leq h - \beta_{N\tau+l\tau+s}^{[\tau]}, & l \in \mathbb{N}_{[0,\hat{N}-1]}, s \in \mathbb{N}_{[1,\tau]}, \\ g^{\mathrm{T}} \Phi_{s-1} \Phi_\tau^l z \leq h - \bar{\beta}_s^{[\tau]}, & l \in \mathbb{N}_{\geq \hat{N}}, s \in \mathbb{N}_{[1,\tau]} \end{array} \right\}. \tag{3.30}$$

To remove the consideration of infinite number of constraints in (3.30), using the Theorem 2.3 from [143], there exists a $n^* \in \mathbb{N}_{\geq 1}$ such that (3.30) are ensured through the first $(\hat{N} + n^*)\tau$ constraints. The terminal constraints for $\tau \in \mathbb{N}_{[1,\bar{\tau}]}$ can be defined as

$$\bar{\mathcal{Z}}_f^{[\tau]} := \left\{ z \in \mathbb{R}^{n_x} \;\middle|\; \begin{array}{ll} g^{\mathrm{T}} \Phi_{s-1} \Phi_\tau^l z \leq h - \beta_{N\tau+l\tau+s}^{[\tau]}, & l \in \mathbb{N}_{[0,\hat{N}-1]}, s \in \mathbb{N}_{[1,\tau]}, \\ g^{\mathrm{T}} \Phi_{s-1} \Phi_\tau^l z \leq h - \bar{\beta}_s^{[\tau]}, & l \in \mathbb{N}_{[\hat{N},\hat{N}+n^*]}, s \in \mathbb{N}_{[1,\tau]} \end{array} \right\}. \tag{3.31}$$

3.4 Stochastic self-triggered MPC with adaptive prediction horizon

3.4.1 Optimization problem and algorithm

Given the state $x(k_j)$ at sampling time instant k_j, the reformulation of the prototype MPC optimization problem $\mathbb{P}^{[\tau]}(x(k_j), \mathbf{c}^{[\tau]}(k_j))$ for $\tau \in \mathbb{N}_{[1,\bar{\tau}]}$ is defined as

$$\begin{aligned}
\min_{\{\mathbf{c}^{[\tau]}(k_j)\}} \quad & \tilde{J}(x(k_j), \mathbf{c}^{[\tau]}(k_j), \tau) \\
& \text{for } l \in \mathbb{N}_{[0,N-1]}, s \in \mathbb{N}_{[1,\tau]}, \\
\text{s.t.} \quad & z(l\tau + s|k_j) = \Phi_{s-1} z(l\tau|k_j) + H_B^{[s]} \mathbf{c}^{[\tau]}(l|k_j), \\
& g^{\mathrm{T}} \Phi_{s-1} \Phi_\tau^l z(0|k_j) + g^{\mathrm{T}} \Phi_{s-1} H_{\Phi_\tau}(l) \mathbf{c}^{[\tau]}(k_j) + g^{\mathrm{T}} H_B^{[s]} \mathbf{c}^{[\tau]}(l|k_j) \leq h - \beta_{l\tau+s}^{[\tau]}, \\
& z(N\tau|k_j) \in \bar{\mathcal{Z}}_f^{[\tau]}.
\end{aligned} \tag{3.32}$$

Define $V_{k_j}^{[\tau]}$ and $\mathbf{c}^{*[\tau]}(k_j)$ as the optimal value function and optimal solution to the corresponding problem $\mathbb{P}^{[\tau]}(x(k_j), \mathbf{c}^{[\tau]}(k_j))$. At each sampling time instant k_j, in order to reduce the communication burden, the largest inter-execution time interval τ_j is obtained by solving the following self-triggered MPC problem $\mathbb{P}^{[\text{st}]}(x(k_j))$:

$$\tau_j^* := \quad \arg\max_{\tau \in \mathbb{N}_{[1,\bar{\tau}]}} \tau$$

$$\text{s.t.} \quad \mathcal{F}_{x(k_j)}^{[\tau]} \neq \emptyset, \tag{3.33}$$

$$V_{k_j}^{[\tau]} \leq V_{k_j}^{[\tau_{j-1}^*]} + \alpha e^{-\gamma \tau_{j-1}^*} \eta_{k_{j-1}} + (e^{-\gamma} - e^{-\gamma \tau_{j-1}^*})\mathbb{E}\{\sum_{s=0}^{\tau_{j-1}^*-1} \|A^s w\|_P\},$$

where $0 < \alpha < 1$ is a tuning parameter, $\eta_{k_{j-1}} = \sum_{s=0}^{\tau_{j-1}^*-1}(\|x(k_{j-1} + s)\|_Q^2 + \|u(k_{j-1} + s)\|_R^2)$ and $\mathcal{F}_{x(k_j)}^{[\tau]}$ is the feasible set with respect to the OCP $\mathbb{P}^{[\tau]}(x(k_j), \mathbf{c}^{[\tau]}(k_j))$. The resulting optimal control sequence is denoted as $\mathbf{c}^{*[\tau_j]}(k_j)$ and the stochastic self-triggered MPC algorithm is summarized in Algorithm 2.

Algorithm 2: Stochastic self-triggered MPC with adaptive prediction horizon

---

Offline: Set $k = 0$. Determine the control gains $K_i, i \in \mathbb{N}_{[0,\bar{\tau}]}$. Define the chance constraints reformulation parameters and terminal constraints parameters.

while *Termination conditions not satisfied* **do**

> **Step 1.** Get the system measure $x(k)$;
>
> **Step 2.** Obtain the inter-execution time interval τ^* and optimal control sequence $\mathbf{c}^{*[\tau]}(k)$ by solving problem $\mathbb{P}^{[\text{st}]}(x(k))$;
>
> **Step 3.** Apply control input $u(k + i) = K_i x(k) + c^*(0|k)$ to the system for $i \in \mathbb{N}_{[0,\tau^*-1]}$;
>
> **Step 4.** Set the next sampling time instant as $k = k + \tau^*$, and return to Step 1.

end

The closed-loop system under the Algorithm 2 is given by

$$x(k + 1) = Ax(k) + Bu(k) + w(k), k \in \mathbb{N}_{[k_j, k_{j+1}-1]}$$

$$u(k) = K_{k-k_j} x(k_j) + c^*(k_j), \tag{3.34}$$

$$k_{j+1} = k_j + \tau_j^*, k_0 = 0, j \in \mathbb{N}.$$

and the closed-loop properties for the system are summarized in the following subsection.

Remark 13. *Even though the prediction horizon adapts to changes in τ, the number of decision variables in the MPC optimization problem is fixed for all $\tau \in \mathbb{N}_{[1,\bar{\tau}]}$ due* *to the sparse structure of self-triggering control patterns* (3.8). *The computational complexity will increase as τ increases because τN linear constraints will be imposed on nominal dynamics and terminal state as shown in* (3.19) *and* (3.31).

3.4.2 Closed-loop properties

Theorem 4. *(Recursive feasibility) At sampling time instant k_0, if the self-triggered MPC problem $\mathbb{P}^{[\mathrm{st}]}(x_{k_0})$ is initially feasible, then for any sampling time instant $k_j, j \in \mathbb{N}_{>0}$, the problem $\mathbb{P}^{[\mathrm{st}]}(x(k_j))$ is feasible. Furthermore, it can be ensured chance constraints* (3.2) *are satisfied for $i \in \mathbb{N}_0$.*

Proof. At sampling time instant k_j, let the solution to self-triggered MPC problem $\mathbb{P}^{[\mathrm{st}]}(x(k))$ be τ_j^* and $\mathbf{c}^{*[\tau_j]}(k_j)$. At the next sampling time instant k_{j+1}, define the candidate solution as $\tau_{j+1} = \tau_j^*$ and $\tilde{\mathbf{c}}^{[\tau_{j+1}]}(k_{j+1}) = \{\mathbf{c}^{*[\tau_j]}(1|k_j), \dots, \mathbf{c}^{*[\tau_j]}(N-1|k_j), \mathbf{0}\}$. For $l \in \mathbb{N}_0$ and $s \in \mathbb{N}_{[1,\tau]}$, it holds that

$$
\begin{aligned}
&g^{\mathrm{T}} \Phi_{s-1} \Phi_\tau^l x(N\tau|k_{j+1}) \\
={}&g^{\mathrm{T}} \Phi_{s-1} \Phi_\tau^{l+1} z(N\tau|k_j) + g^{\mathrm{T}} \Phi_{s-1} \Phi_\tau^{l+1} \sum_{n=0}^{s-1} A^{s-1-n} W(k_j + N\tau + n) \\
\leq{}&h - \beta_{(l+1)\tau+s}^{[\tau]} + d_{(l+1)\tau+s}^{[\tau]} \leq h - \beta_{l\tau+s}^{[\tau]}
\end{aligned}
\tag{3.35}
$$

Therefore we can obtain that $z(N\tau|k_{j+1}) \in \bar{\mathcal{Z}}_f^{[\tau_j]}$. From Theorem 3 and terminal constraint satisfaction, it can be concluded that at sampling time instant k_{j+1}, the candidate solution $\tilde{\mathbf{c}}^{[\tau_{j+1}]}(k_{j+1})$ is a feasible solution to the self-triggered problem $\mathbb{P}^{[\mathrm{st}]}(x(j+1))$, and by induction, the optimization problem is feasible at all sampling time instant $k_j, j \in \mathbb{N}_0$. $\qquad\square$

Theorem 5. *(Stability) Consider the closed-loop system* (3.34) *under Algorithm 2, we have*

$$
\lim_{n \to \infty} \frac{1}{n} \sum_{j=0}^{n} \sum_{k=0}^{\tau_j^*-1} \mathbb{E}(\|x(k)\|_Q^2 + \|u(k)\|_R^2) \leq \frac{e^{-\gamma}}{(1-\alpha)e^{-\gamma\bar{\tau}}} \mathbb{E}\{\sum_{n=0}^{\bar{\tau}-1} \|A^s w\|_P\}.
\tag{3.36}
$$

Proof. At sampling time instant k_j, denote the optimal solution to the self-triggered problem $\mathbb{P}^{[\mathrm{st}]}(x(k_j))$ as τ_j^* and $\mathbf{c}^{[\tau_j^*]}(k_j) = \{\mathbf{c}^{[\tau_j^*]}(0|k_j), \dots, \mathbf{c}^{[\tau_j^*]}(N-1|k_j)\}$. At the next sampling time instant k_{j+1}, define the candidate solution $\tilde{\mathbf{c}}^{[\tau_j^*]}(k_{j+1})$ with respect to inter-execution time τ_j^* as $\tilde{\mathbf{c}}^{[\tau_j^*]}(k_{j+1}) = \{\mathbf{c}^{[\tau_j^*]}(1|k_j), \ldots, \mathbf{c}^{[\tau_j^*]}(N-1|k_j), \mathbf{0}\}$. The self-triggering condition and optimality of $\mathbb{P}^{[\text{st}]}(x(k_{j+1}))$ at sampling time k_{j+1} implies that

$$
\begin{aligned}
&\mathbb{E}\left\{V_{k_{j+1}}^{[\tau_{j+1}^*]}(x(k_{j+1}), \mathbf{c}^{[\tau_{j+1}^*]}(k_{j+1}))\right\} - \mathbb{E}\left\{V_{k_j}^{[\tau_j^*]}(x_{k_j}, \mathbf{c}^{[\tau_j^*]}(k_j))\right\} \\
&\leq \mathbb{E}\left\{V_{k_{j+1}}^{[\tau_j^*]}(x(k_{j+1}), \mathbf{c}^{[\tau_j^*]}(k_{j+1}))\right\} - \mathbb{E}\left\{V_{k_j}^{[\tau_j^*]}(x_{k_j}, \mathbf{c}^{[\tau_j^*]}(k_j))\right\} + \alpha e^{-\gamma \tau_j^*} \eta_{k_j} \\
&\quad + (e^{-\gamma} - e^{-\gamma \tau_j^*})\mathbb{E}\{\sum_{s=0}^{\tau_j^*-1} \|A^s w\|_P\} \\
&\leq \mathbb{E}\left\{\tilde{V}_{k_{j+1}}^{[\tau_j^*]}(x(k_{j+1}), \tilde{\mathbf{c}}^{[\tau_j^*]}(k_{j+1}))\right\} - \mathbb{E}\left\{V_{k_j}^{[\tau_j^*]}(x_{k_j}, \mathbf{c}^{[\tau_j^*]}(k_j))\right\} + \alpha e^{-\gamma \tau_j^*} \eta_{k_j} \\
&\quad + (e^{-\gamma} - e^{-\gamma \tau_j^*})\mathbb{E}\{\sum_{s=0}^{\tau_j^*-1} \|A^s w\|_P\}.
\end{aligned}
$$

where $\eta_{k_j} = \sum_{s=0}^{\tau_j^*-1}(\|x_{k_j+s}\|_Q^2 + \|u_{k_j+s}\|_R^2)$. It should be noted that from the feasibility analysis in Theorem 4, $\mathbb{P}^{[\text{st}]}(x(k_{j+1}))$ admits a solution with inter-execution time τ_j^*. From (3.11), it holds that $x(k_{j+1}) = z(0|k_{j+1}) + e(0|k_{j+1})$, where $z(0|k_{j+1}) = z(\tau_j^*|k_j) = \Phi_{\tau_j^*} x(k_j) + H_B^{[\tau_j^*]} \mathbf{c}^{[\tau_j^*]}(0|k_j)$, and $e(0|k_{j+1}) = \sum_{n=0}^{\tau_j^*-1} A^{\tau_j^*-1-n} w(k_j + n)$. Therefore,

$$
\begin{aligned}
&\mathbb{E}\left\{\tilde{V}_{k_{j+1}}^{[\tau_j^*]}(x(k_{j+1}), \tilde{\mathbf{c}}^{[\tau_j^*]}(k_{j+1}))\right\} - \mathbb{E}\left\{V_{k_j}^{[\tau_j^*]}(x(k_j), \mathbf{c}^{[\tau_j^*]}(k_j))\right\} \\
&= -e^{-\gamma \tau_j^*} \sum_{s=0}^{\tau_j^*-1} \mathbb{E}\{\|x(s|k_j)\|_Q^2 + \|u(s|k_j)\|_R^2\} \\
&\quad + e^{-\gamma \tau_j^*} \mathbb{E}\{\sum_{l=0}^{N-1} \|\Phi_{\tau_j^*}^l e(0|k_{j+1})\|_{Q^{[\tau_j^*-1]}}^2 + \|\Phi_{\tau_j^*}^N e(0|k_{j+1})\|_P^2\},
\end{aligned}
$$

where the last term is equal to $\mathbb{E}\{\|e(0|k_{j+1})\|_P^2\}$ from (2). So it implies that

$$
\begin{aligned}
&\mathbb{E}\left\{V_{k_{j+1}}^{[\tau_{j+1}^*]}(x(k_{j+1}), \mathbf{c}^{[\tau_{j+1}^*]}(k_{j+1}))\right\} - \mathbb{E}\left\{V_{k_j}^{[\tau_j^*]}(x(k_j), \mathbf{c}^{[\tau_j^*]}(k_j))\right\} \\
&\leq -(1-\alpha)e^{-\gamma \tau_j^*} \sum_{s=0}^{\tau_j^*-1}(\|x(k_j + s)\|_Q^2 + \|u(k_j + s)\|_R^2) + e^{-\gamma}\mathbb{E}\{\|e(0|k_{j+1})\|_P^2\} \\
&\leq -(1-\alpha)e^{-\gamma \bar{\tau}} \sum_{s=0}^{\bar{\tau}_j^*-1}(\|x(k_j + s)\|_Q^2 + \|u(k_j + s)\|_R^2) + e^{-\gamma}\mathbb{E}\{\sum_{s=0}^{\bar{\tau}-1} \|A^s w\|_P\}
\end{aligned}
\tag{3.37}
$$

Summing (3.37) from $j = 0$ to $j = n$ results in

$$\mathbb{E}\{V_{k_0}^{[\tau_0^*]}\} - \mathbb{E}\{V_{k_n}^{[\tau_n^*]}\}$$
$$\geq (1-\alpha)e^{-\gamma\bar{\tau}}\sum_{j=0}^{n}\sum_{s=0}^{\tau_j^*-1}(\|x(k_j+s)\|_Q^2 + \|u(k_j+s)\|_R^2) - ne^{-\gamma}\mathbb{E}\{\sum_{s=0}^{\bar{\tau}-1}\|A^s w\|_P\},$$

impling the quadratic stability condition (3.36) since $\mathbb{E}\{V_{k_n}^{[\tau_n^*]}\} - \mathbb{E}\{V_{k_0}^{[\tau_0^*]}\}$ is bounded.

$\square$

3.5 Numerical examples

Consider the linearized DC-DC converter system as shown in [121] and [110]:

$$x(k+1) = \begin{bmatrix} 1 & 0.0075 \\ -0.143 & 0.996 \end{bmatrix} x(k) + \begin{bmatrix} 4.798 \\ 0.115 \end{bmatrix} u(k) + w(k),$$
$$\Pr\left\{\begin{bmatrix} 1 & 0 \end{bmatrix} x(k) \leq 2\right\} \geq 0.8, \tag{3.38}$$

where $w(k)$ is assumed to be a truncated identical independently distributed Gaussian random process with zero mean and a variance 0.04^2. $w(k)$ is bounded by $\|w(k)\| \leq 0.1$, and $Q = \text{diag}\{1, 10\}$, $R = 1$. The prediction horizon is chosen as $N = 10$, the extended horizon is chosen as $\hat{N} = 5$, the n^* in (3.31) is selected as 1, and the maximal triggering interval is selected as $\bar{\tau} = 5$. For $s \in \mathbb{N}_{[0,\bar{\tau}-1]}$, the state feedback gains K_s in (3.7) are given in Table 3.1, and it can be readily verified that all Φ_s are Schur stable. The terminal weighting matrix P is selected by (3.23) and Remark 11

Table 3.1: Selection of feedback gain K_s.

	Feedback gain K_s	
K_0	$[-0.2093$	$0.0766]$
K_1	$[5.6403 \times 10^{-4}$	$-0.0349]$
K_2	$[9.8827 \times 10^{-5}$	$-0.0142]$
K_3	$[1.3912 \times 10^{-4}$	$-0.0084]$
K_4	$[1.8643 \times 10^{-4}$	$-0.0059]$

as $P = \begin{bmatrix} 24.1653 & -102.5252; & -102.5252 & 605.0106 \end{bmatrix}$. The tuning parameter α in self-tiggered condition (3.33) and the damping parameter γ are chosen as $\alpha = 0.01$ and $\gamma = 0.001$, respectively. Simulation studies are provided to demonstrate the effectiveness of the proposed method in comparison with the self-triggered SMPC control scheme with a fixed prediction horizon in [110]. For each control strategy, $N_{trial} = 100$ realizations of uncertainty sequence are generated with initial condition $\begin{bmatrix} 2.5 & 2.8 \end{bmatrix}^{\mathrm{T}}$ and the simulation length is $T_{sim} = 20$ steps.

Chance constraint violations: The state trajectories of the closed-loop system are illustrated in the left plot in Figure 3.2, where the alternative blue and red lines denote the evolution of state trajectory with respect to time steps. The black line denotes the chance constraint, and the right plot in Figure 3.2 enlarges the constraint bound region to show constraint violations. With the selected parameters, it can be observed that probabilities of constraint violations at time step $1, 3, 5, 7$ are 19.3%, 19.6%, 19.8% and 20.3%, respectively. The simulation results implies that the proposed self-triggered SMPC controller steers the closed-loop trajectories to the region around origin while the constraint violations probability is tight to the specific value 20%.

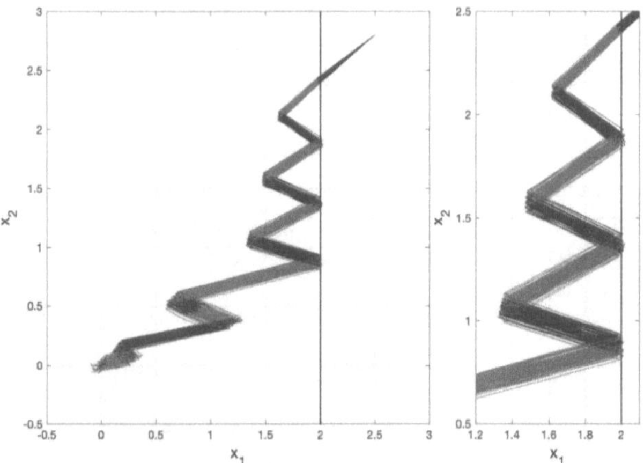

Figure 3.2: State trajectories of closed-loop system under self-triggered MPC with adaptive prediction horizon.

Average inter-execution time and performance: The closed-loop trajectories of state x_1, control input u and inter-execution interval under the self-triggered SMPC algorithm with an adaptive prediction horizon (STSMPC-AP) and a fixed prediction horizon (STSMPC-FP) for one realization of uncertainty sequence are illustrated in Figure 3.3. The sampling instants are highlighted by red circles for STSMPC-AP and blue diamonds for STSMPC-FP to demonstrate the different sampling behaviour between the two algorithms. Considering 100 realizations of uncertainties, the average inter-execution time for STSMPC-AP is 2.67 while it is 1.81 for STSMPC-FP. As shown in the bottom plot in Figure 3.3, it can be observed that the inter-execution time converges to the maximal triggering length $\bar{\tau} = 5$ for the proposed STSMPC-AP while that almost converges to 1 for STSMPC-FP. This difference arises from the improved design of triggering condition (3.33).

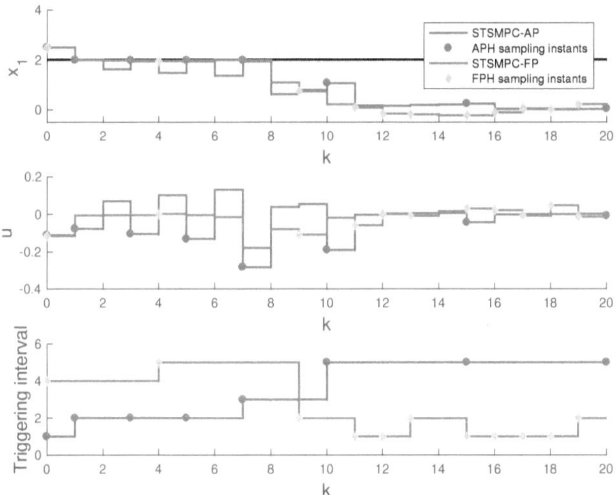

Figure 3.3: State trajectories, control inputs and triggering interval of closed-loop system under STSMPC-AP and STSMPC-FP with one realization of uncertainty sequence.

To further evaluate the proposed algorithm, the performance index is defined as

$$J_{index} = \frac{1}{N_{trial}} \frac{1}{T_{sim}} \sum_{k=0}^{T_{sim}} (\|x(k)\|_Q^2 + \|u(k)\|_R^2). \tag{3.39}$$

It can be obtained that the measure is 15.9292 for STSMPC-AP and 15.5632 for STSMPC-FP. Compared with the self-triggered SMPC method with a fixed prediction horizon, our proposed algorithm achieves a more desirable asymptotic sampling behaviour without sacrificing the performance too much. This can also be demonstrated from the top plot in Figure 3.3.

Impacts of tuning parameters on average sampling interval: To analyze the impact of tuning parameters γ and α on the proposed algorithm, 50 and 10 different values of γ and α are evenly chosen in the intervals $[10^{-3}, 10^{-1}]$ and $[0.01, 0.4]$. The relationship between the average sampling interval with respect to γ and α are plotted in Figure 3.4. It can be observed that the average sampling interval increases as γ and α increase.

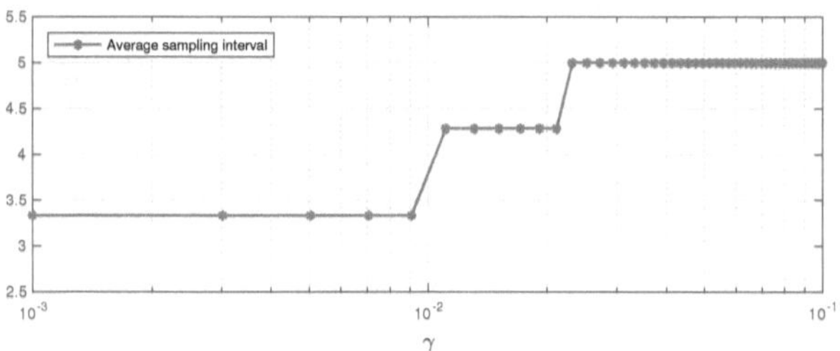

(a) The average sampling interval versus different value of γ. $\alpha = 0.01$ is selected as a constant.

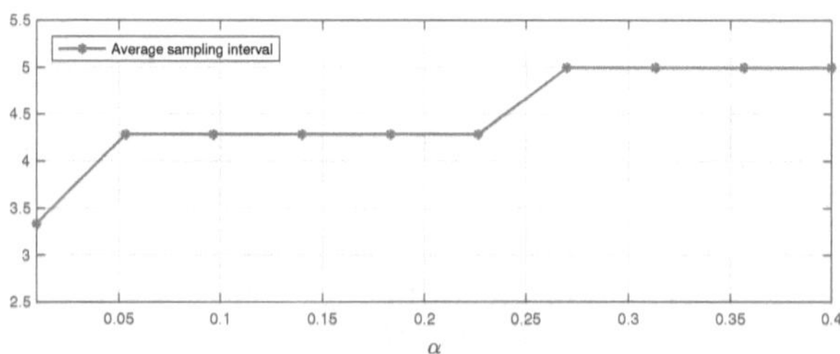

(b) The average sampling interval versus different value of α. $\gamma = 0.001$ is selected as a constant.

Figure 3.4: The average sampling interval versus tuning parameters γ and α.

3.6 Conclusions

In this chapter, a novel self-triggered SMPC algorithm with an adaptive prediction horizon is proposed for linear systems subject to both additive disturbances and state chance constraints. The prediction horizon in the MPC algorithm changes adaptively to generate some appropriate inter-execution time intervals. To deal with the additive disturbance, an improved triggering condition is designed, and the asymptotic sampling behaviour is analyzed. Sufficient conditions to guarantee the recursive feasibility of the algorithm are given, and the closed-loop system is proven to be quadratically stable. Simulation results have shown the efficacy of the designed self-triggered control method in reducing the communication burden while guaranteeing some specific performance loss.

Chapter 4

Distributed Self-triggered Stochastic MPC for CPSs with Coupled Chance Constraints: A Stochastic Tube Approach

4.1 Introduction

With the development of computer technologies and communication networks, cyber-physical systems (CPSs) have become an interest of research due to the comprehensive integration of physically engineered systems, such as sensors, actuators and plants, with intricate cyber components, possessing information communication and computation. In CPSs, advantages of low installation cost, high reliability, flexible modularity, improved efficiency, and greater autonomy can be obtained by the tight coordination of physical and cyber components. Several sectors, including robotics, transportation, health care, smart building, and smart grid, have witnessed the successful application of CPSs design. The model-based control synthesis plays a vital role in CPSs design as the dynamic behaviour can be systematically adjusted under this approach. Due to the heterogeneous and spatially interconnected nature of CPSs, it necessitates the adoption of distributed control structure to improve the structural flexibility and scalability while maintaining some desirable closed-loop properties. Meanwhile, the integration of extensive cyber capability and physical plants with ubiquitous uncertainties also introduces concerns over the robustness and stability of the CPSs. Thus, in order to achieve satisfactory performance, robustness and stability, a detailed investigation into control synthesis of CPSs under the distributed paradigm is of importance.

The distributed controller design for CPSs is also contingent on physical constraints and performance considerations and distributed MPC (DMPC) has attracted much attention since it simultaneously handles the need for meeting system constraints and the quest for desirable performance goals. In the literature, DMPC has been extensively studied and applied for large-scale CPSs, such as nonlinear chemical systems [144], natural gas refrigeration plant [145] and unicycle robots [146]. For distributed networked systems subject to uncertainties with a given probability distribution, stochastic DMPC has been investigated over the last several years. For subsystems that are dynamically decoupled [88], the Chebyshev type inequality is used to transform the coupled chance constraints into deterministic form. In [93], for distributed systems subject to parametric uncertainties, the uncertainty propagation is approximated by the generalized polynomial chaos expansions (gPCEs), and a gPCEs-based DMPC method with guaranteed closed-loop properties is proposed. The tube-based stochastic MPC in [43] is extended to the distributed form in [90] for a linear system subject to additive disturbances. The DMPC control strategy in [90] is then further extended for systems with both parametric and additive uncertainties, where the stochastic tube technique in [69] is utilized to deal with coupled constraints.

It is well known that the handling of coupling in DMPC relies on the subsystem update rule design. In the literature, the sequential and iterative update rule have been proposed, which will lead to a heavy communication burden for the network. It is therefore interesting to study the event-based distributed MPC, which will reduce the communication burden for the network. Interested readers please refer to [100, 136] for a detailed review of event-based control, involving event-triggered control and self-triggered control. Several results [137, 114, 138] on self-triggered MPC have been developed for systems without uncertainties in the literature over the last few years. For the system under bounded uncertainties, results on robust self-triggered MPC can be found in [132, 113]. In [140], a robust self-triggered SMPC control method is proposed using the reachability analysis. Furthermore, [111] and [110] extend the result [113] to the stochastic setting where the system is affected by stochastic disturbances. In [111], the Cantelli inequality is utilized to construct the tightened constraints, while in [110, 112] the constraints are constructed using the probability distribution explicitly following the ideas in [43]. It should be noted that self-triggered

DMPC schemes receive relatively little attention and only reported in [147, 148, 149, 142]. In [147, 148, 149], the self-triggered mechanism and DMPC algorithm are designed separately, while in [142], the self-triggered mechanism and DMPC are co-designed following the line of [138].

Motivated by the aforementioned discussions, a distributed self-triggered SMPC control strategy is proposed for linear CPSs subject to additive stochastic disturbance and coupled chance constraints. One important class of controlled systems in stochastic DMPC design is that all subsystems are dynamically decoupled but share coupled constraints. The aim of this work is to extend the self-triggered SMPC framework described in [110] to the distributed paradigm by making the following modifications: (i) deterministic reformulations of both local and coupled chance constraints are constructed to evaluate the uncertainty propagation through the distributed systems; (ii) a sequential self-triggered update rule is designed to achieve a tradeoff between the overall system performance and communication among each subsystem; (iii) to construct recursive feasible stochastic tubes, terminal constraints are redesigned considering both coupled chance constraints and inter-execution time. Contributions of this work are given as follows:

- Both local and coupled chance constraints are handled in a cooperative fashion by using the distribution information of uncertainties arising from either local subsystem or other neighbouring subsystems;

- The amount of communication among each subsystem and computation required by each subsystem are significantly reduced thanks to the co-design of the self-triggered mechanism and distributed SMPC controller, while the sacrifice of overall system performance can be tuned to some specific level of tradeoff;

- Sufficient conditions on constraint parameters tightening are developed to guarantee the recursive feasibility of the algorithm, and the quadratic stability of the overall system is investigated in the presence of additive stochastic disturbances.

The remainder of this chapter is organized as follows. Section 4.2 presents the problem setup of the distributed CPSs subject to coupled chance constraints and reviews the self-triggered SMPC scheme in [110] for a single system. This then ushers in the centralized self-triggered SMPC method presented in Section 4.3 for distributed systems subject to coupled chance constraints. Beginning from Section 4.4, we seek to formulate the distributed self-triggered SMPC algorithm and establish sufficient conditions to guarantee recursive feasibility and stability. Section 4.5 compares the performance of the proposed algorithm with the corresponding distributed SMPC method by numerical examples. The chapter concludes in Section 4.6 with a concise summary.

Notations: In the following, sets of natural numbers and real numbers are denoted as $\mathbb{N}$ and $\mathbb{R}$. For any $a, b \in \mathbb{N}$, define $\mathbb{N}_0$, $\mathbb{N}_{\geq a}$, $\mathbb{N}_{\leq a}$, $\mathbb{N}_{[a,b]}$ as sets $\{n \in \mathbb{N} | n \geq 0\}$, $\{n \in \mathbb{N} | n \geq a\}$, $\{n \in \mathbb{N} | n \leq a\}$, $\{n \in \mathbb{N} | b \leq n \leq a\}$, respectively. $x(k)$ denotes state at time k, and $x(i|k)$ denotes predicted i-step ahead state given the state $x(k)$. For a random variable x, denote $\Pr\{x\}$ and $\mathbb{E}(x)$ as the probability and expectation of x, respectively. For $n \in \mathbb{N}_{\geq 1}$, the matrix $I_{n \times n}$ denotes the identity matrix in $\mathbb{R}^{n \times n}$.

4.2 Problem formulation

4.2.1 Distributed cyber-physical control systems

The studied distributed CPSs, as shown in Figure 4.1, feature that subsystems together with associated actuators and sensors are spatially separated and connected via a communication network. The system state measurements are communicated from the sensors to the online MPC controller; then, the next sampling time instant and generated control sequences are transmitted to the actuator through the same network. Consider the following cyber-physical control systems consisting of N_p subsystems:

$$x_p(k+1) = A_p x_p(k) + B_p u_p(k) + D_p w_p(k), p \in \mathcal{P} := \mathbb{N}_{[1,N_p]}, \tag{4.1}$$

in which $x_p(k) \in \mathbb{R}^{n_{p,x}}$, $u_p(k) \in \mathbb{R}^{n_{p,u}}$ and $w_p(k) \in \mathbb{R}^{n_{p,w}}$ denote system state, control input, and stochastic additive disturbance for subsystem p with appropriate system matrices A_p, B_p and D_p. The additive disturbance sequence $\{w_p(0), w_p(1), \dots\}$ is assumed to be independent and identically distributed with zero mean and covariance matrix $\sigma^{w_p} \in \mathbb{R}^{n_{p,w} \times n_{p,w}}$. For $w_p(k) = [w_{p,1}(k) \dots w_{p,n_w}(k)]^{\mathrm{T}}$, the distribution of $w_{p,i}(k)$ is given by

$$\Pr\{w_{p,i}(k) \leq \xi_{p,i}\} = \begin{cases} 1 & , \ \xi_{p,i} \geq \alpha_{p,i}, \\ F_{p,i}(\xi_{p,i}) & , \ -\alpha_{p,i} \leq \xi_{p,i} \leq \alpha_{p,i}, \\ 0 & , \ \xi_{p,i} \leq -\alpha_{p,i}, \end{cases} \tag{4.2}$$

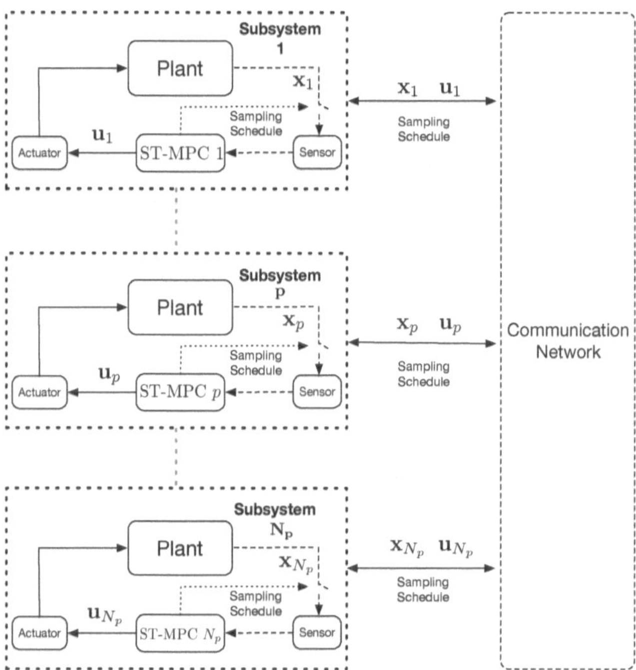

Figure 4.1: System configuration of distributed self-triggered CPSs with N_p subsystems.

where $\alpha_{p,i} \in \mathbb{R}_{>0}$ and it follows that $w_p(k)$ is assumed to lie in a polytope $\mathbb{W}_p = \{w_p : |w_p| \leq \alpha_p\}$ with $\alpha_p = [\alpha_{p,1} \ldots \alpha_{p,n_w}]$. The sequence $\{w_p(0), \ldots, w_p(k), \ldots\}$ is a realization of random process $W_p(k), k \in \mathbb{N}_0$.

Each subsystem p is assumed to be subject to both local chance constraints and coupled chance constraints in the form of

$$\Pr\left\{g_p^{\mathrm{T}} x_p(i+1|k) \leq h_p\right\} \geq p_p, \qquad i \in \mathbb{N}_0, p \in \mathcal{P}, \qquad (4.3a)$$

$$\Pr\left\{\sum_{p=1}^{N_p} g_{cp}^{\mathrm{T}} x_p(i+1|k) \leq h_c\right\} \geq p_{p,c}, \qquad i \in \mathbb{N}_0, c \in \mathcal{C}, \qquad (4.3b)$$

in which vectors g_p, h_p describe the local chance constraints for subsystem $p \in \mathcal{P}$ and vectors g_{cp}, h_c, $p \in \mathcal{P}, c \in \mathcal{C} := \mathbb{N}_{[1,N_c]}$ characterize the coupling between each subsystem, where N_c denotes the number of coupled constraints. Parameters p_p and $p_{p,c}$ represent probabilities of local and coupled constraints violation, respectively. Define $\mathcal{P}_c := \{p \in \mathcal{P}|g_{cp} \neq 0\}$ as the set of subsystems involved in the coupled constraint c and $\mathcal{C}_p := \{c \in \mathcal{C}|g_{cp} \neq 0\}$ as the set of coupled constraints involved in subsystem p. The set $\mathcal{Q}_p := (\cup_{c \in \mathcal{C}}\mathcal{P}_c)\backslash\{p\}$ is defined as all coupled subsystems to subsystem p.

4.2.2 Self-triggered mechanism overview

As suggested in [150], at each sampling instant, only one subsystem is permitted to update the control sequence by solving the DMPC problem. Figure 4.2(a) shows the periodical sampling behaviour for a group consisting of three subsystems, and the dashed line denotes the communication between each subsystem. To reduce the communication burden between each subsystem, the self-triggered mechanism is implemented on the cyber-physical system (4.1), as shown in Figure 4.2(b).

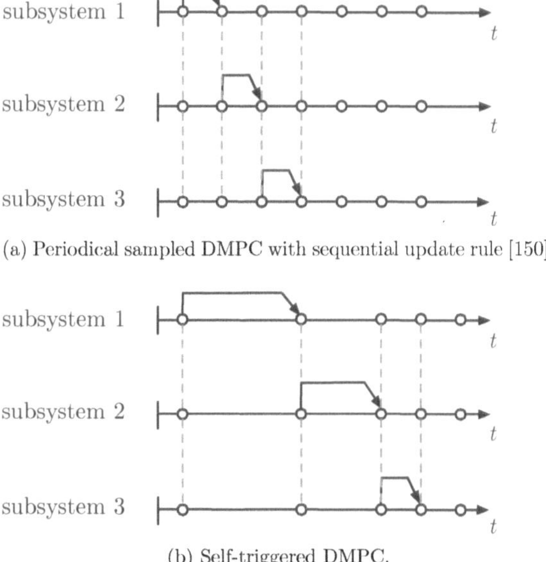

(a) Periodical sampled DMPC with sequential update rule [150].

(b) Self-triggered DMPC.

Figure 4.2: Illustration of periodical and self-triggered sampling scheme for CPSs with 3 subsystems.

Different from the periodically-triggered distributed SMPC scheme, the states $x_p(k)$ are only measured and transmitted to neighbor subsystems at sampling time instant $k_j, j \in N_0$, where the triggering time sequence is defined as $\{k_0, \ldots, k_j, \ldots\}$ with $k_{j+1} = k_j + \tau_j^*$ and $k_0 = 0$. In the following, preliminaries on self-triggered SMPC problem formulation described in [110] for a single subsystem are reviewed. The optimized variable $\tau_j^* \in N_{\geq 1}$ is the inter-execution time interval, which is determined by the self-triggered mechanism at sampling time instant k_j. Define an integer $\tau \in N_{[1,\bar{\tau}]}$ as the fixed sampling interval where $\bar{\tau} \in N_{[1,N-1]}$ is the offline given maximum of inter-execution time interval and N is the prediction horizon. At sampling time instant k_j and for subsystem p, the predicted control input $u_p(i|k_j;\tau)$ corresponding to τ can be parameterized as

$$
u_p(i|k_j;\tau) = \begin{cases} K_p z_p(i|k_j) + v_p(i|k_j;\tau), \; i \in N_{[0,\tau-1]}, & (4.4a) \\ K_p x_p(i|k_j) + v_p(i|k_j;\tau), i \in N_{[\tau,N-1]}, & (4.4b) \\ K_p x_p(i|k_j), & i \in N_{\geq N}, & (4.4c) \end{cases}
$$

where $z_p(i|k_j)$ denotes the predicted nominal state, with $z_p(0|k_j) = x_p(k_j)$. For each subsystem p, the feedback gain matrix $K_p \in \mathbb{R}^{n_{p,x} \times n_{p,u}}$ is chosen offline such that the matrix $\Phi_p = A_p + B_p K_p$ is Schur stable. The variables $v_p(i|k_j;\tau), i \in N_{[0,N-1]}$ are control perturbations in the prediction associated to inter-execution time interval τ. Define $\mathbf{v}_p(k_j;\tau) = \left[v_p^{\mathrm{T}}(0|k_j;\tau); \ldots; v_p^{\mathrm{T}}(N-1|k_j;\tau) \right]^{\mathrm{T}}$ as the matrix of decision variables, which are determined by solving the MPC optimization problem defined below. Note that, the predicted control input $u_p(i|k_j;\tau)$ for first τ step depend only on $x_p(k_j), v_p(i|k_j;\tau)$ and is therefore deterministic. Due to the implementation of the self-triggered mechanism, there is no communication between the controller and sensor during the period $[k_j, k_j + \tau]$, and the subsystem p are therefore controlled in an open-loop fashion. Meanwhile, the disturbed state feedback (4.4b) is introduced after τ steps to reduce the effect of disturbance in prediction.

Inspired by [48] and [110], the local cost function at sampling time instant k_j for subsystem p corresponding to $\tau \in N_{[1,\bar{\tau}]}$ is defined as the expected value of an infinite horizon quadratic cost

$$J_{O,p}(\mathbf{v}_p(k_j;\tau)) = \frac{1}{\alpha}\sum_{i=0}^{\tau-1}\mathbb{E}(\|x_p(i|k_j)\|^2_{Q_p} + \|u_p(i|k_j;\tau)\|^2_{R_p})$$
$$+ \sum_{i=\tau}^{\infty}\mathbb{E}(\|x_p(i|k_j)\|^2_{Q_p} + \|u_p(i|k_j;\tau)\|^2_{R_p}). \tag{4.5}$$

where $Q_p \succ 0$ and $R_p \succeq 0$ are weighting matrices and $\alpha \geq 1$ is a tuning parameter penalizing the open-loop phase arising from the self-trigger mechanism. Therein, the prototype MPC problem $\mathbf{P}^{[\tau]}_{O,p}(\mathbf{v}_p(k_j;\tau))$ for subsystem p with a fixed sampling interval τ is formulated as

$$
\begin{aligned}
\min_{\mathbf{v}_p(k_j;\tau)} \quad & J_{O,p}(\mathbf{v}_p(k_j;\tau)) \\
\text{s.t.} \quad & x_p(i+1|k) = A_p x_p(i|k) + B_p u_p(i|k;\tau) + D_p w_p(k+i), \quad i \in \mathbb{N}_0, \\
& \Pr\left\{g_p^{\mathrm{T}} x_p(i+1|k) \leq h_p\right\} \geq p_p, \quad i \in \mathbb{N}_0, \\
& \Pr\left\{\textstyle\sum_{p=1}^{N_p} g_{cp}^{\mathrm{T}} x_p(i+1|k) \leq h_c\right\} \geq P_{p,c}, \quad i \in \mathbb{N}_0.
\end{aligned}
\tag{4.6}
$$

The feasible set to the optimization problem $\mathbf{P}^{[\tau]}_{O,p}(\mathbf{v}_p(k_j;\tau))$ is defined as

$$\mathcal{F}^{[\tau]}_{O,p}(x_p(k_j)) := \{\mathbf{v}_p(k_j;\tau)|\mathbf{P}^{[\tau]}_{O,p}(\mathbf{v}_p(k_j;\tau)) \text{ feasible}\}. \tag{4.7}$$

Define the optimal value function of $\mathbf{P}^{[\tau]}_{O,p}(\mathbf{v}_p(k_j;\tau))$ as $V^{[\tau]}_{O,p}(x(k_j)) := J_{O,p}(\mathbf{v}^*_p(k_j;\tau))$, where $\mathbf{v}^*_p(k_j;\tau) := \arg\min_{\mathbf{v}_p(k_j;\tau)} J_{O,p}(\mathbf{v}_p(k_j;\tau))$ denotes the optimal solution to the optimization problem $\mathbf{P}^{[\tau]}_{O,p}(\mathbf{v}_p(k_j;\tau))$. The prototype self-triggered problem $\mathbf{P}^{\mathrm{ST}}_{O,p}(x(k_j))$ for subsystem p at sampling time instant k_j is therefore defined as

$$\tau^*_j := \max\left\{\tau \in \mathbb{N}_{[1,\bar{\tau}]} \;\middle|\; \begin{array}{l} \mathcal{F}^{[\tau]}_{O,p}(x_p(k_j)) \neq \emptyset \\ V^{[\tau]}_{O,p}(x(k_j)) \leq V^{[1]}_{O,p}(x(k_j)) \end{array} \right\}, \tag{4.8}$$
$$\mathbf{v}^*_p(k_j) = \mathbf{v}^*_p(k_j;\tau^*_j).$$

For subsystem p, by solving $\mathbf{P}^{\mathrm{ST}}_{O,p}(x(k_j))$ at time instant k_j, the next sampling time instant k_{j+1} and the control input during the period $[k_j, k_{j+1}]$ are given by $k_{j+1} = $

$k_j + \tau_j^*$ and $\mathbf{v}_p^*(k_j)$. The resulting closed-loop system is then formulated as

$$
\begin{aligned}
x_p(k+1) &= A_p x_p(k) + B_p u_p(k) + D_p w(k), \\
u_p(k) &= K_p z_p(i|k) + v_p^*(i|k), k \in \mathbb{N}_{[k_j, k_{j+1}-1]}, \\
k_{j+1} &= k_j + \tau_j^*.
\end{aligned}
\tag{4.9}
$$

For a single subsystem p controlled by the self-triggered SMPC scheme, the closed-loop properties of (4.9) have been well discussed in [110]. However, when subsystems share coupled chance constraints, the existing self-triggered mechanism cannot be applied directly. Meanwhile, the infinite horizon cost and infinite number of chance constraints make the prototype optimization problem $\mathbf{P}_{O,p}^{[\tau]}(\mathbf{v}_p(k_j; \tau))$ intractable for implementation, leading to the quest of some appropriate approximations and modifications to the MPC problem $\mathbf{P}_{O,p}^{[\tau]}(\mathbf{v}_p(k_j; \tau))$. Therefore, the aim of this work is twofold: (i) to design a computationally tractable MPC formulation of $\mathbf{P}_{O,p}^{[\tau]}(\mathbf{v}_p(k_j; \tau))$ for online implementation; (ii) to desgin a distributed self-triggered mechanism for the overall system to reduce the communication burden between each subsystem while guranteeing the closed-loop properties.

4.3 Centralized Self-triggered SMPC problem under coupled chance constraints

In this section, we will extend the self-triggered SMPC control strategy proposed in [110] to the CPSs (4.1) subject to coupled chance constraints in a centralized fashion. To be more specific, the reformulation of the cost function and chance constraints handling will be presented. Meanwhile, the designed constraints tightening parameters will be utilized in the definition of a distributed algorithm for chance constraints satisfaction. Then, improved terminal constraints will be introduced to guarantee the recursive feasibility of the centralized algorithm. Finally, discussions about the closed-loop properties of the centralized self-triggered SMPC strategy will be given. The centralized algorithm will not only provide a benchmark to evaluate the distributed algorithm performance but also plays an important role in the initialization of the distributed paradigm.

4.3.1 Cost function reformulation

For each subsystem $p \in \mathcal{P}$ with given τ, the control parametrization defined in (4.4) is utilized and the predicted state $x_p(i|k)$ can be decomposed into nominal state part $z_p(i|k_j) = \mathbb{E}[x_p(i|k_j)]$ and error state part $e_p(i|k_j)$, as shown below:

$$x_p(i|k_j) = z_p(i|k_j) + e_p(i|k_j), \tag{4.10}$$

where

$$z_p(i|k_j) = \Phi_p^i z_p(0|k_j) + H_p^{[i]} \mathbf{v}_p(k_j; \tau),$$

$$e_p(i|k_j) = \begin{cases} \displaystyle\sum_{l=0}^{i-1} A_p^{i-l-1} D_p W_p(l|k_j), & i \in \mathbb{N}_{[1,\tau]}, \\ \displaystyle \Phi_p^{i-\tau} e_p(\tau|k_j) + \sum_{l=0}^{i-\tau-1} \Phi_p^{i-l-1} D_p W_p(l|k_j), & i \in \mathbb{N}_{\geq \tau+1}, \end{cases} \tag{4.11}$$

in which the initial error state $e_p(0|k_j)$ is assumed to be $e_p(0|k_j) = 0$, and the matrix $H_p^{[i]}$ is defined as $H_p^{[i]} = \begin{bmatrix} \Phi_p^{i-1} B_p & \cdots & B_p & \mathbf{0} \end{bmatrix}$. Inspired by [48], an autonomous description of the predicted dynamics for $i \in \mathbb{N}_{\geq \tau}$ can be generated by

$$\eta_p(i+1|k_j) = \Psi_p \eta_p(i|k_j) + \delta_p(k_j + i), i \in \mathbb{N}_{\geq \tau}, \tag{4.12}$$

where

$$\eta_p(\tau|k_j) = \begin{bmatrix} x_p(\tau|k_j)^{\mathrm{T}} & (M_p^\tau \mathbf{v}_p(k_j; \tau))^{\mathrm{T}} \end{bmatrix}^{\mathrm{T}},$$

$$\Psi_p = \begin{bmatrix} \Phi_p & B_p E_p \\ 0 & M_p \end{bmatrix}, \delta_p(k_j + i) = \begin{bmatrix} w_p(k_j + i) \\ 0 \end{bmatrix},$$

$$E_p = \begin{bmatrix} I_{n_{p,u} \times n_{p,u}} & 0 & \cdots & 0 \end{bmatrix},$$

$$M_p = \begin{bmatrix} 0 & I_{n_{p,u} \times n_{p,u}} & 0 & \cdots & 0 \\ 0 & 0 & I_{n_{p,u} \times n_{p,u}} & \cdots & 0 \\ \vdots & \vdots & & & \vdots \\ 0 & 0 & 0 & \cdots & 0 \end{bmatrix}.$$

The variable $\eta_p(i|k_j)$ is defined as the i-step ahead augmented state given $x_p(k_j)$ and $\mathbf{v}_p(k_j; \tau)$ at k_j. The difference between the autonomous model (4.12) and that in [48] is that we use the τ-step ahead predicted state $x_p(\tau|k_j) = z_p(\tau|k_j) + e_p(\tau|k_j)$ here to construct the initial augmented state in (4.12). So the error between two time instants k_j and $k_j + \tau$ is taken into account to derive the convergence of augmented state $\lim\limits_{i\to\infty} \eta_p(i|k_j)$, which is given by the following proposition.

Proposition 3. *(Convergence of $\eta_p(i+1|k_j), i \in \mathbb{N}_{\geq\tau}$) Given $\tau \in \mathbb{N}_{[1,\bar{\tau}]}$, the sequence $\{\eta_p(i+i|k_j)\}, i \in \mathbb{N}_{\geq\tau}$, generated by (4.12) satisfies $\lim\limits_{i\to\infty} \mathbb{E}\{\eta_p(i|k_j)\} = 0$ and $\lim\limits_{i\to\infty} \mathbb{E}\{\eta_p(i|k_j)\eta_p(i|k_j)^{\mathrm{T}}\} = \Theta_p$, where Θ_p is the solution to Lyapunov equation*

$$\Theta_p - \Psi_p\Theta_p\Psi_p^{\mathrm{T}} = \delta_p(k_j+i)\delta_p^{\mathrm{T}}(k_j+i), \tag{4.13}$$

if and only if there exists $P_p \succ 0$ satisfying

$$P_p - \Psi_p P_p \Psi_p^{T} \succ 0. \tag{4.14}$$

Proof. Define the nominal and error state of $\eta_p(i|k_j)$ as $\bar{\eta}_p(i|k_j) = \mathbb{E}\{\eta_p(i|k_j)\}$ and $\eta_{e,p}(i|k_j) = \eta_p(i|k_j) - \bar{\eta}_p(i|k_j)$, respectively. For $i \in \mathbb{N}_{\geq\tau}$, it holds that

$$\bar{\eta}_p(i+1|k_j) = \Psi_p\bar{\eta}_p(i|k_j), \tag{4.15a}$$

$$\eta_{e,p}(i+1|k_j) = \Psi_p\eta_{e,p}(i|k_j) + \delta_p(k_j+i), \tag{4.15b}$$

with $\eta_{e,p}(\tau|k_j) = \begin{bmatrix} e_p^{\mathrm{T}}(\tau|k_j) & 0 \end{bmatrix}^{\mathrm{T}}$. It follows directly from the mean square stable (MSS) condition (4.14) and (4.15a) that $\lim\limits_{i\to\infty} \bar{\eta}_p(i|k_j) = 0$.

Define $\hat{\Theta}_p(i|k_j) = \mathbb{E}\{\eta_{e,p}(i|k_j)\eta_{e,p}^{\mathrm{T}}(i|k_j)\} - \Theta_p$. From (4.15b), we have

$$\mathbb{E}\{\eta_{e,p}(i+1|k_j)\eta_{e,p}^{\mathrm{T}}(i+1|k_j)\}$$
$$=\Psi_p\mathbb{E}\{\eta_{e,p}(i|k_j)\eta_{e,p}^{\mathrm{T}}(i|k_j)\}\Psi_p^{\mathrm{T}} + \delta_p(k_j+i)\delta_p^{\mathrm{T}}(k_j+i).$$

Following (4.13), it implies that

$$\begin{aligned}
\hat{\Theta}_p(i+1|k_j) &= \mathbb{E}\{\eta_{e,p}(i+1|k_j)\eta_{e,p}^{\mathrm{T}}(i+1|k_j)\} - \Theta_p \\
&= \Psi_p\mathbb{E}\{\eta_{e,p}(i|k_j)\eta_{e,p}^{\mathrm{T}}(i|k_j)\}\Psi_p^{\mathrm{T}} + \delta_p(k_j+i)\delta_p^{\mathrm{T}}(k_j+i) - \Theta_p \\
&= \Psi_p\left(\mathbb{E}\{\eta_{e,p}(i|k_j)\eta_{e,p}^{\mathrm{T}}(i|k_j)\} - \Theta_p\right)\Psi_p^{\mathrm{T}} \\
&= \Psi_p\hat{\Theta}_p(i|k_j)\Psi_p^{\mathrm{T}}.
\end{aligned}$$

From the MSS condition in (4.14), it follows that $\lim\limits_{i\to\infty} \hat{\Theta}_p(i|k_j) = 0$, and hence

$$\lim\limits_{i\to\infty} \mathbb{E}\{\eta_{e,p}(i|k_j)\eta_{e,p}^{\mathrm{T}}(i|k_j)\} = \Theta_p.$$

Thus the proof is complete. $\qquad\square$

Remark 14. *As shown in Proposition 3, the covariance matrix $\mathbb{E}\{\eta_p(i|k_j)\eta_p(i|k_j)^T\}$ will converge to a fixed finite value Θ_p as $i \to \infty$ regardless of the selection of the inter-execution time τ. As shown in (4.4b), periodical state feedback is introduced to the control parametrization for $i \in \mathbb{N}_{\geq\tau}$ and Φ_p is designed to be schur stable, so the initial error $\eta_{e,p}(\tau|k_j)$ will be eliminated as $i \to \infty$.*

Hence, the sum of predicted cost of all subsystems over $i \in \mathbb{N}_{\geq\tau}$ can be expressed as

$$\sum_{p=1}^{N_p}\sum_{i=\tau}^{\infty}\mathbb{E}(\|x_p(i|k_j)\|^2_{Q_p} + \|u_p(i|k_j)\|^2_{R_p}) = \sum_{p=1}^{N_p}\sum_{i=\tau}^{\infty}\mathbb{E}(\eta_p^{\mathrm{T}}(i|k_j)\tilde{Q}_p\eta_p(i|k_j)), \qquad (4.16)$$

where the augmented state weighting matrix $\tilde{Q}_p$ is defined as

$$\tilde{Q}_p = \begin{bmatrix} Q_p + K_p^{\mathrm{T}}R_pK_p & K_p^{\mathrm{T}}R_pE_p \\ E_p^{\mathrm{T}}R_pK_p & E_p^{\mathrm{T}}R_pE_p \end{bmatrix}.$$

By Proposition 3, the stage cost in (4.16) converges to a finite value

$$L_p := \sum_{p=1}^{N_p}\lim_{i\to\infty}\mathbb{E}(\eta_p^{\mathrm{T}}(i|k_j)\tilde{Q}_p\eta_p(i|k_j)) = \mathrm{tr}(\Theta_p\tilde{Q}_p)$$

along trajectories of (4.12), so the predicted cost in (4.16) is infinite. Hence, to obtain a finite cost, the centralized cost function for the whole system with $\tau \in \mathbb{N}_{[1,\bar{\tau}]}$ is defined as

$$J_C(\mathbf{v}_C(k_j;\tau)) = \frac{1}{\alpha}\sum_{p=1}^{N_p}\sum_{i=0}^{\tau-1}\mathbb{E}(\|x_p(i|k_j)\|^2_{Q_p} + \|u_p(i|k_j;\tau)\|^2_{R_p} - L_p)$$

$$+ \sum_{p=1}^{N_p}\sum_{i=\tau}^{\infty}\mathbb{E}(\eta_p^{\mathrm{T}}(i|k_j)\tilde{Q}_p\eta_p(i|k_j) - L_p), \qquad (4.17)$$

where the matrix $\mathbf{v}_C(k_j;\tau) = \begin{bmatrix} \mathbf{v}_1(k_j;\tau); \ldots; \mathbf{v}_{N_p}(k_j;\tau) \end{bmatrix}$ contains control perturbation variables corresponding to τ, and $\mathbf{v}_p(k_j;\tau)$ for subsystem p are defined in the Section 4.2.2. The centralized cost function (4.17) is intractable for online implementation because it involves an infinite horizon cost. The cost function defined in (4.17) can be rewritten as a quadratic form of decision variables $\mathbf{v}(k_j)$ as shown in the following lemma.

Lemma 4. *The second summation term in* (4.17) *evaluated along* (4.12) *is given by*

$$\sum_{p=1}^{N_p}\sum_{i=\tau}^{\infty}\mathbb{E}(\eta_p^{\mathrm{T}}(i|k_j)\tilde{Q}_p\eta_p(i|k_j) - L_p) = \sum_{p=1}^{N_p}\mathbb{E}\left\{\begin{bmatrix}\eta_p(\tau|k_j)\\1\end{bmatrix}^{\mathrm{T}}\tilde{P}_p\begin{bmatrix}\eta_p(\tau|k_j)\\1\end{bmatrix}\right\}$$
$$=\sum_{p=1}^{N_p}\begin{bmatrix}\bar{\eta}_p(\tau|k_j)\\1\end{bmatrix}^{\mathrm{T}}\tilde{P}_p\begin{bmatrix}\bar{\eta}_p(\tau|k_j)\\1\end{bmatrix} + \mathrm{tr}(\tilde{P}_p\theta_{p,w}^{[\tau]}),$$

(4.18)

where the matrices $\tilde{P}_p$ and $\theta_{p,w}^{[\tau]}$ are defined as

$$\tilde{P}_p = \begin{bmatrix}P_{p,z} & 0\\0 & P_{p,c}\end{bmatrix},\ \theta_{p,w}^{[\tau]} = \begin{bmatrix}\sum_{s=0}^{\tau-1}A_p^s D_p\sigma^{w_p}(A_p^s D_p)^{\mathrm{T}} & \cdots & \cdots\\ & \vdots & & 0\\ & \vdots & & 1\end{bmatrix},$$

with $P_{p,z}$ and $P_{p,c}$ are given by

$$P_{p,z} - \Psi_p^{\mathrm{T}}P_{p,z}\Psi_p = \tilde{Q}_p, \tag{4.19a}$$
$$P_{p,c} = -\mathrm{tr}(\Theta P_{p,z}). \tag{4.19b}$$

Proof. Define a function $V_p(i|k_j) = \eta_p^{\mathrm{T}}(i|k_j)P_{p,z}\eta_p(i|k_j) + P_{p,c}$ for $i \geq \tau$. From (4.12), it holds that

$$\mathbb{E}(V_p(i|k_j)) - \mathbb{E}(V_p(i+1|k_j))$$
$$=\mathbb{E}(\eta_p^{\mathrm{T}}(i|k_j)P_{p,z}\eta_p(i|k_j)) - \mathbb{E}(\eta_p^{\mathrm{T}}(i+1|k_j)P_{p,z}\eta_p(i+1|k_j)) \tag{4.20}$$
$$=\mathbb{E}\left(\eta_p^{\mathrm{T}}(i|k_j)(P_{p,z} - \Psi_p^{\mathrm{T}}P_{p,z}\Psi_p)\eta_p(i|k_j)\right) - \mathbb{E}(\delta_p^{\mathrm{T}}(k_j+i)P_{p,z}\delta_p(k_j+i)).$$

From (4.19a), it holds that

$$\mathbb{E}\left(\eta_p^{\mathrm{T}}(i|k_j)(P_{p,z} - \Psi_p^{\mathrm{T}}P_{p,z}\Psi_p)\eta_p(i|k_j)\right) = \mathbb{E}\left(\eta_p^{\mathrm{T}}(i|k_j)\tilde{Q}_p\eta_p(i|k_j)\right). \tag{4.21}$$

By post-multiplying $P_{p,z}$ and extracting the trace, (4.13) becomes

$$\mathbb{E}\{\delta_p^{\mathrm{T}}(k_j+i)P_{p,z}\delta_p(k_j+i)\} =\mathrm{tr}(\Theta_p P_{p,z} - \Psi_p\Theta_p\Psi_p^{\mathrm{T}}P_{p,z})$$
$$=\mathrm{tr}(\Theta_p(P_{p,z} - \Psi_p^{\mathrm{T}}P_{p,z}\Psi_p)) = \mathrm{tr}(\Theta_p\tilde{Q}_p).$$

(4.22)

Hence, from (4.21) and (4.22), (4.20) can be expressed as

$$\mathbb{E}(V_p(i|k_j)) - \mathbb{E}(V_p(i+1|k_j)) = \mathbb{E}\left(\eta_p^{\mathrm{T}}(i|k_j)\tilde{Q}_p\eta_p(i|k_j)\right) - \mathrm{tr}(\Theta_p\tilde{Q}_p).$$

For $i \in \mathbb{N}_{\geq\tau}$, summing the above recursion over $p \in \mathbb{N}_{[1,N_p]}$ gives

$$\sum_{p=1}^{N_p}(\mathbb{E}(V_p(\tau|k_j)) - \lim_{i\to\infty}\mathbb{E}(V_p(i|k_j))) = \sum_{p=1}^{N_p}\sum_{i=\tau}^{\infty}\mathbb{E}(\|x_p(i|k_j)\|_{Q_p}^2 + \|u_p(i|k_j)\|_{R_p}^2 - L_p).$$

From the definition of (4.19b) and $V_p(i|k_j)$, it implies that $\lim_{i\to\infty}\mathbb{E}(V_p(i|k_j)) = 0$. Therefore, the first equality of (4.18) is verified. Due to the linearity of (4.12), the second equality can be readily verified since $\bar{\eta}_p(i|k_j) = \mathbb{E}\{\eta_p(i|k_j)\}$ and $\eta_{e,p}(\tau|k_j) = \left[e_p^{\mathrm{T}}(\tau|k_j)\ \ 0\right]^{\mathrm{T}}$, with $e_p(\tau|k_j) = \sum_{s=0}^{\tau-1}A_p^{\tau-1-s}D_p w_p(k_j+s)$. Thus the proof is complete. $\square$

Consequently, by Lemma 4 and (4.4), a computationally tractable reformulation of the cost function (4.17) is given by

$$\begin{aligned}
J_C(\mathbf{v}_C(k_j;\tau)) =& \frac{1}{\alpha}\sum_{p=1}^{N_p}\sum_{i=0}^{\tau-1}\left(\|z_p(i|k_j)\|_{Q_p+K_p^{\mathrm{T}}R_pK_p}^2 + \|v_p(i|k_j;\tau)\|_{R_p}^2\right. \\
&+ 2z_p(i|k_j)^{\mathrm{T}}R_p v_p(i|k_j;\tau) + \sum_{l=0}^{i-1}\mathrm{tr}(Q_p A_p^l D_p \sigma^{w_p}(A_p^l D_p)^{\mathrm{T}}) - L_p\Bigg) \\
&+ \sum_{p=1}^{N_p}\left(\begin{bmatrix}\bar{\eta}_p(\tau|k_j)\\1\end{bmatrix}^{\mathrm{T}}\tilde{P}_p\begin{bmatrix}\bar{\eta}_p(\tau|k_j)\\1\end{bmatrix} + \mathrm{tr}(\tilde{P}_p\theta_{p,w}^{[\tau]})\right).
\end{aligned}$$

$$(4.23)$$

4.3.2 Local and coupled chance constraints handling

To ensure chance constraints satisfactions (4.3) in the closed-loop operation, necessary and sufficient conditions are provided in following lemmas, which transforms chance constraints into deterministic forms.

Lemma 5. *(Local and coupling chance constraints handling) Given $\tau \in \mathbb{N}_{[1,\bar{\tau}]}$, for each subsystem p, local chance constraints $\Pr\left\{g_p^T x_p(i|k_j) \leq h_p\right\} \geq p_p$ and coupled chance constraints $\Pr\left\{\sum_{p=1}^{N_p} g_{cp}^T x_p(i|k_j) \leq h_c\right\} \geq P_{p,c}$ are satisfied if and only if there* *exists* $\mathbf{v}_p(k_j; \tau)$ *such that*

$$g_p^T z_p(i|k_j) \leq h_p - \gamma_{p,i}^{[\tau]}, i \in \mathbb{N}_{\geq 1}, \tag{4.24a}$$

$$\sum_{p=1}^{N_p} g_{cp}^T z_p(i|k_j) \leq h_c - \nu_{c,i}^{[\tau]}, i \in \mathbb{N}_{\geq 1}, \tag{4.24b}$$

where $\gamma_{p,i}^{[\tau]}$ *is defined as the minimum value such that*

$$\Pr\left\{ g_p^T \sum_{l=0}^{i-1} A_p^{i-l-1} D_p W_p(l|k_j) \leq \gamma_{p,i}^{[\tau]} \right\} = p_p, i \in \mathbb{N}_{[1,\tau]}, \tag{4.25a}$$

$$\Pr\left\{ g_p^T \left(\Phi_p^{i-\tau} e_p(\tau|k_j) + \sum_{l=0}^{i-\tau-1} \Phi_p^{i-l-1} D_p W_p(l|k_j) \right) \leq \gamma_{p,i}^{[\tau]} \right\} = p_p, i \in \mathbb{N}_{\geq \tau+1}. \tag{4.25b}$$

and $\nu_{c,i}^{[\tau]}$ *is the minimum value such that*

$$\Pr\left\{ \sum_{p=1}^{N_p} g_{cp}^T \sum_{l=0}^{i-1} A_p^{i-l-1} D_p W_p(l|k_j) \leq \nu_{c,i}^{[\tau]} \right\} = p_{p,c}, i \in \mathbb{N}_{[1,\tau]}, \tag{4.26a}$$

$$\Pr\left\{ \sum_{p=1}^{N_p} g_{cp}^T \left(\Phi_p^{i-\tau} e_p(\tau|k_j) + \sum_{l=0}^{i-\tau-1} \Phi_p^{i-l-1} D_p W_p(l|k_j) \right) \leq \nu_{c,i}^{[\tau]} \right\} = p_{p,c}, i \in \mathbb{N}_{\geq \tau+1}. \tag{4.26b}$$

Proof. Conditions for local chance constraints satisfaction (4.24a) have been given in Lemma 3.1 in [110]. Considering the coupled chance constraints (4.3b), we have

$$\Pr\left\{ \sum_{p=1}^{N_p} g_{cp}^T x_p(i|k_j) \right\} = \Pr\left\{ \sum_{p=1}^{N_p} g_{cp}^T z_p(i|k_j) + \sum_{p=1}^{N_p} g_{cp}^T e_p(i|k_j) \right\} \leq h_c, i \in \mathbb{N}_{\geq 1},$$

where $\sum_{p=1}^{N_p} g_{cp}^T e_p(i|k_j)$ represents the sum of stochastic components of the prediction of $\sum_{p=1}^{N_p} g_{cp}^T x_p(i|k_j)$. From (4.26), $\nu_{c,i}^{[\tau]}$ is the minimum value such that the probability of the sum of stochastic components is greater than $\nu_{c,i}^{[\tau]}$ is $p_{p,c}$. Therefore, (4.24b) is equivalent to (4.3b) for $i \in \mathbb{N}_{\geq 1}$. $\qquad \square$

Remark 15. *The computation of parameters* $\gamma_{p,i}^{[\tau]}$ *and* $\nu_{c,i}^{[\tau]}$ *relies on either numerically approximation to the relevant probability distribution [151] or sampling-based approximation [112]. The essential assumption of conducting this approximation is that* w_p *is distributed independently for all* $p \in \mathcal{P}$.

Similar to Theorem 3.1 in [110], recursive feasible constraint tightening conditions can be derived by modifying Lemma 5 as:

Lemma 6. *(Recursive feasible constraint tightening) Define matrix $\Gamma_p^{[\tau]}$ and matrix $\Gamma_c^{[\tau]}$ as shown in (4.27) and (4.28).*

$$\Gamma_p^{[\tau]} = \begin{bmatrix} \gamma_{p,1}^{[\tau]} & \cdots & \gamma_{p,\tau}^{[\tau]} & \gamma_{p,\tau+1}^{[\tau]} & \gamma_{p,\tau+2}^{[\tau]} & \cdots \\ 0 & \cdots & 0 & b_{p,\tau+1}^{[\tau]} + \xi_{p,\tau+1}^{[\tau]} & b_{p,\tau+2}^{[\tau]} + \xi_{p,\tau+2}^{[\tau]} & \cdots \\ 0 & \cdots & 0 & 0 & b_{p,\tau+2}^{[\tau]} + d_{p,\tau+2}^{[\tau]} + \xi_{p,\tau+1}^{[\tau]} & \cdots \\ \vdots & \vdots & \vdots & \vdots & \vdots & \ddots \end{bmatrix}, \qquad (4.27)$$

where $d_{p,i}^{[\tau]} = \max\limits_{w \in \mathbb{W}_p} g_p^T \Phi^{i-\tau-1} w$, $b_{p,i}^{[\tau]} = \max\limits_{w \in \mathbb{W}_p} g_p^T \Phi_p^{i-\tau} \sum\limits_{\tau-1}^{l} A_p^l w$ and $\xi_{p,i}^{[\tau]}$ is the minimum value such that $\Pr\left\{ \sum\limits_{l=0}^{i-\tau-1} g_p^T \Phi_p^l w_p \leq \xi_{p,i}^{[\tau]} \right\} = p_p$,

$$\Gamma_c^{[\tau]} = \begin{bmatrix} \nu_{c,1}^{[\tau]} & \cdots & \nu_{c,\tau}^{[\tau]} & \nu_{c,\tau+1}^{[\tau]} & \nu_{c,\tau+2}^{[\tau]} & \cdots \\ 0 & \cdots & 0 & b_{c,\tau+1}^{[\tau]} + \xi_{c,\tau+1}^{[\tau]} & b_{c,\tau+2}^{[\tau]} + \xi_{c,\tau+2}^{[\tau]} & \cdots \\ 0 & \cdots & 0 & 0 & b_{c,\tau+2}^{[\tau]} + d_{c,\tau+2}^{[\tau]} + \xi_{c,\tau+1}^{[\tau]} & \cdots \\ \vdots & \vdots & \vdots & \vdots & \vdots & \ddots \end{bmatrix}, \qquad (4.28)$$

where $d_{c,i}^{[\tau]} = \sum\limits_{p=1}^{N_p} g_{cp}^T \max\limits_{w \in \mathbb{W}_p} \Phi_p^{i-\tau-1} D_p w$, $b_{c,i}^{[\tau]} = \sum\limits_{p=1}^{N_p} g_{cp}^T \Phi_p^{i-\tau} D_p \max\limits_{w \in \mathbb{W}_p} \sum\limits_{\tau-1}^{l} A_p^l D_p w$ and $\xi_{c,i}^{[\tau]}$ is the minimum value such that $\Pr\left\{ \sum\limits_{p=1}^{N_p} g_{cp}^T \sum\limits_{l=0}^{i-\tau-1} \Phi_p^l D_p w_p \leq \xi_{c,i}^{[\tau]} \right\} = p_c$. At time instant k_j, if there exists $\mathbf{v}_p(k_j; \tau)$ satisfying:

$$g_p^T z_p(i|k_j) \leq h_p - \beta_{p,i}^{[\tau]}, i \in \mathbb{N}_{\geq 1}, \qquad (4.29a)$$

$$\sum_{p=1}^{N_p} g_{cp} z_p(i|k_j) \leq h_c - \zeta_{c,i}^{[\tau]}, i \in \mathbb{N}_{\geq 1}, \qquad (4.29b)$$

where $\beta_{p,i}^{[\tau]}$ and $\zeta_{c,i}^{[\tau]}$ are the maximum element of the ith column of matrix $\Gamma_p^{[\tau]}$ and $\Gamma_c^{[\tau]}$. Then there exists at least one solution at time instant k_{j+1} such that local and coupled chance constraints defined in (4.3) are satisfied for all $i \in \mathbb{N}_{\geq 1}$.

Proof. The existense and recursive feasibility of local probabilitic constraint at the next sampling time instant k_{j+1} have been presented in Theorem 3.1 in [110] and details are omitted here. In the following, we consider the construction of recursive feasible tube for the coupled constraints. If $\zeta_{c,i}^{[\tau]}$ is defined as the first row in (4.28), then (4.29b) is equivalent to (4.24b). At time instant k_{j+1} with $\tau = 1$, the candidate solution for subsystem $p \in \mathcal{P}$ can be defined as $\tilde{\mathbf{v}}_p(k_{j+1}; 1) = M_p^{[\tau]} \mathbf{v}_p(k_j; \tau) = \left[v_p^{\mathrm{T}}(\tau|k_j; \tau); \ldots; v_p^{\mathrm{T}}(N-1|k_j; \tau); \mathbf{0}; \ldots; \mathbf{0} \right]^{\mathrm{T}}$. By (4.10), it follows that for $i \in \mathbb{N}_0$,

$$x_p(i|k_{j+1}) = \Phi_p^{\tau+i} x_p(k_j) + H_p^{[\tau+i]} \mathbf{v}_p(k_j; \tau) + \Phi_p^i e_p(0|k_{j+1}) + \sum_{l=0}^{i-1} \Phi_p^{i-1-l} D_p w_p(l|k_{j+1}),$$

$$= z_p(i+\tau|k_j) + \Phi_p^i e_p(0|k_{j+1}) + \sum_{l=0}^{i-1} \Phi_p^{i-1-l} D_p w_p(l|k_{j+1}),$$

where $e_p(0|k_{j+1}) = \sum_{i=0}^{\tau-1} A_p^{\tau-1-i} D_p w_p(i|k_j)$ denotes the error between two sampling time instant k_j and k_{j+1}. To ensure the existence of a feasible solution at time k_{j+1}, the worst-case realization of $e_p(k_{j+1}|k_{j+1})$ is considered, and hence the coupled constraints (4.3b) can be reformulated as

$$\sum_{p=1}^{N_p} g_{cp}^{\mathrm{T}} z_p(i|k_j) \le h_c - b_{c,i}^{[\tau]} - \xi_{c,i}^{[\tau]}, i \in \mathbb{N}_{\ge \tau+1},$$

which follows the second row in (4.28). Similarly, the predicted state at sampling time instant $k_{j+l}, l \ge 2$, with $\tau = 1$ follows

$$x_p(i|k_{j+l}) = z_p(\tau + l + i|k_{j+l}) + \Phi_p^i e_p(0|k_{j+l}) + \sum_{l=0}^{i-1} \Phi_p^{i-1-l} D_p w_p(l|k_{j+l}), i \in \mathbb{N}_0,$$

where $e_p(0|k_{j+l}) = \Phi_p^{l-\tau} D_p \sum_{i=0}^{\tau-1} A_p^{\tau-1-i} D_p w_p(i|k_j) + \sum_{s=\tau+1}^{l+\tau} \Phi_p^{s-\tau-1} D_p w_p(\tau+s|k_j)$. So the feasibility of the solution at sampling time instant $k_{j+l}, l \ge 2$ can be ensured by

$$\sum_{p=1}^{N_p} g_{cp}^{\mathrm{T}} z_p(i|k_j) \le h_c - b_{c,i}^{[\tau]} - \sum_{s=0}^{\tau-2} d_{c,i-s}^{[\tau]} - \xi_{c,i-l+1}^{[\tau]}, i \in \mathbb{N}_{\ge \tau+l}.$$

Therefore, the existense of a feasible solution at sampling time instant $k_{j+l}, l \in \mathbb{N}_0$ can be ensured if $\zeta_{c,i}^{[\tau]}$ is selected as the maximum element of the ith column of (4.28). $\quad\square$

Remark 16. *The proof of Lemma 6 is an extension of Theorem 3.1 in [110] where the* *coupled chance constraints are considered. The parameters $\beta_{p,i}^{[\tau]}$ and $\zeta_{c,i}^{[\tau]}$ are determined by the prediction step i and inter-execution time τ.*

To guarantee the chance constraints are satisfied over an infinite prediction horizon, the terminal set should be constructed. The terminal dynamics of the nominal system can be written as $z_p(N + i + 1|k_j) = \Phi_p z_p(N + i|k_j), i \in \mathbb{N}_0$, and terminal constraints are imposed to the nominal state $z_p(i|k_j), i \in \mathbb{N}_{\geq N}$ to deal with the infinite prediction horizon. For each subsystem $p \in \mathcal{P}$, the terminal constraints $\mathcal{Z}_{p,f}^{[\tau]}$ for dealing with local chance constraints (4.3a) are defined as

$$\mathcal{Z}_{p,f}^{[\tau]} := \left\{ z \in \mathbb{R}^{n_{p,x}} \;\middle|\; g_p^{\mathrm{T}} \Phi_p^i z \leq h_p - \beta_{p,N+i}^{[\tau]}, i \in \mathbb{N}_0 \right\}, \tag{4.30}$$

and similarly, the terminal constraints for the coupled chance constraints (4.3b) are given by

$$\sum_{p=1}^{N_p} g_{cp}^{\mathrm{T}} \Phi_p^i z_p(N|k_j) \leq h_c - \zeta_{c,N+i}^{[\tau]}, i \in \mathbb{N}_0. \tag{4.31}$$

Bounds for $\beta_{p,N+i}^{[\tau]}$ and $\zeta_{p,N+i}^{[\tau]}$ with $i \in \mathbb{N}_{\geq \tau+1}$ are given by the following lemma.

Lemma 7. *For $\tau \in \mathbb{N}_{[1,\bar{\tau}]}$ and $p \in \mathcal{P}$, there exist positive scalars $0 < \rho_p < 1$ and positive definite matrices S_p such that the sequences $\beta_{p,N+i}^{[\tau]}$ and $\zeta_{c,N+i}^{[\tau]}$ for $i \in \mathbb{N}_{\geq \tau+1}$ is upper bounded by*

$$\beta_{p,N+i}^{[\tau]} \leq \bar{\beta}_p^{[\tau]} := \bar{b}_p^{[\tau]} + \sum_{l=\tau+2}^{v_p-1} d_{p,l}^{[\tau]} + \frac{\rho_p^{v_p}}{1-\rho_p}\|g_p\|_{S_p} + \gamma_{p,1}^{[\tau]}, \tag{4.32a}$$

$$\zeta_{c,N+i}^{[\tau]} \leq \bar{\zeta}_c^{[\tau]} := \bar{b}_c^{[\tau]} + \sum_{l=\tau+2}^{v_c-1} d_{c,l}^{[\tau]} + \sum_{p=1}^{N_p} \frac{\rho_p^{v_c}}{1-\rho_p}\|g_{cp}\|_{S_p} + \nu_{c,1}^{[\tau]}, \tag{4.32b}$$

with integers $v_p, v_c \in \mathbb{N}_{\geq 3}$. The bounds on $b_{p,i}^{[\tau]}$ and $b_{c,i}^{[\tau]}$ are given by

$$\bar{b}_p^{[\tau]} := \max_{i \in \mathbb{N}_{\geq \tau+1}, w \in \mathbb{W}} g_p^{T} \Phi_p^{i-\tau} D_p \sum_{l=0}^{\tau-1} A_p^l D_p w,$$

$$\bar{b}_c^{[\tau]} := \max_{i \in \mathbb{N}_{\geq \tau+1}, w \in \mathbb{W}} \sum_{i=1}^{N_p} g_{cp} \Phi_p^{i-\tau} D_p \sum_{l=0}^{\tau-1} A_p^l D_p w.$$

Proof. The bounds (4.32a) on parameters $\beta_{p,i}^{[\tau]}$ have been proved in Lemma 3.4 in [110], so it is omitted here. For the coupled constraints, the existence of the bound $\bar{b}_c^{[\tau]}$ can be guaranteed since Φ_p are all strictly stable. For $i \in \mathbb{N}_{[1,\tau]}$, $\zeta_{c,i}^{[\tau]} = \nu_{c,i}^{[\tau]}$ holds from (4.28). For $i \in \mathbb{N}_{\geq\tau+1}$, it can be readily verified that $\nu_{c,i}^{[\tau]} \leq b_{c,i}^{[\tau]} + \xi_{c,i}^{[\tau]}$ and $\xi_{c,i}^{[\tau]} \leq d_{c,i}^{[\tau]} + \xi_{c,i-1}^{[\tau]}$ because the worst-case realization consideration in the RHS of the inequalities. Hence, it holds by iteration that $\zeta_{c,i}^{[\tau]} = b_{c,i}^{[\tau]} + \sum_{s=M+2}^{i} d_{c,s}^{[\tau]} + \nu_{c,1}^{[\tau]}$ for $i \in \mathbb{N}_{\geq\tau+1}$. In addition, it holds that $\sum_{s=M+2}^{i} d_{c,s}^{[\tau]} \leq \sum_{s=M+2}^{\infty} d_{c,s}^{[\tau]}$. Following the Corollary 4 in [151], the bound on $\sum_{s=M+2}^{\infty} d_{c,s}^{[\tau]}$ can be calculated by $\sum_{l=\tau+2}^{v_c-1} d_{c,l}^{[\tau]} + \sum_{p=1}^{N_p} \frac{\rho_p^{v_c}}{1-\rho_p} \|g_{cp}\|_{S_p}$, leading to (4.32b). $\qquad\square$

By using Lemma 7, we can split the terminal prediction horizon into two parts: $i \leq \tau + \hat{N}$ and $i \geq \tau + \hat{N} + 1$. Following Theorem 2 in [152], the terminal constraint $\mathcal{Z}_{p,f}^{[\tau]}$ and (4.31) can be approximated by

$$\hat{\mathcal{Z}}_{p,f}^{[\tau]} := \left\{ z \in \mathbb{R}^{n_{p,x}} \;\middle|\; \begin{array}{ll} g_p^{\mathrm{T}} \Phi_p^i z \leq h - \beta_{p,N+i}^{[\tau]}, & i \in \mathbb{N}_{[0,\tau+\hat{N}]}, \\ g_p^{\mathrm{T}} \Phi_p^i z \leq h - \bar{\beta}_p^{[\tau]}, & i \in \mathbb{N}_{[\tau+\hat{N}+1,\tau+\hat{N}+n^*]} \end{array} \right\}, \qquad (4.33)$$

and

$$\begin{aligned}
\sum_{p=1}^{N_p} g_{cp}^{\mathrm{T}} \Phi_p^i z_p(N|k_j) &\leq h - \zeta_{p,N+i}^{[\tau]}, i \in \mathbb{N}_{[0,\tau+\hat{N}]}, \\
\sum_{p=1}^{N_p} g_{cp}^{\mathrm{T}} \Phi_p^i z_p(N|k_j) &\leq h - \bar{\zeta}_p^{[\tau]}, i \in \mathbb{N}_{[\tau+\hat{N}+1,\tau+\hat{N}+n^*]},
\end{aligned} \qquad (4.34)$$

where n^* is the smallest integer such that the infinite number of constraints can be ensured through the first $\tau + \hat{N} + n^*$ constraints.

4.3.3 Centralized self-triggered SMPC algorithm

At sampling time instant k_j, define the augmented state $\mathbf{x}_c(k_j)$ for the overall system as $\mathbf{x}_c(k_j) = \left[x_1^{\mathrm{T}}(k_j), \ldots, x_p^{\mathrm{T}}(k_j), \ldots, x_{N_p}^{\mathrm{T}}(k_j) \right]^{\mathrm{T}}$, with $p \in \mathcal{P}$. Then the centralized optimization problem $\mathbf{P}_C^{[\tau]}(\mathbf{x}_c(k_j))$ for the prototype optimization problem (4.6) with a fixed $\tau \in \mathbb{N}_{[1,\bar{\tau}]}$ is formulated as

$$
\begin{aligned}
\min_{\mathbf{v}_C(k_j;\tau)} \quad & J_C(\mathbf{v}_C(k_j;\tau)) \\
& \text{for } p \in \mathcal{P}, c \in \mathcal{C}, \\
\text{s.t.} \quad & z_p(0|k_j) = x_p(k_j), \\
& z_p(i+1|k_j) = \Phi_p z_p(i|k_j) + B_p v_p(i|k_j;\tau), \quad i \in \mathbb{N}_{[0,N-1]} \\
& g_p^{\mathrm{T}} z_p(i|k_j) \le h_p - \beta_{p,i}^{[\tau]}, \quad i \in \mathbb{N}_{[1,N-1]}, \\
& z_p(N|k_j) \in \hat{\mathcal{Z}}_{p,f}^{[\tau]}, \\
& \sum_{p=1}^{N_p} g_{cp} z_p(i|k_j) \le h_c - \eta_{c,i}^{[\tau]}, \quad i \in \mathbb{N}_{[1,N-1]}, \\
& \sum_{p=1}^{N_p} g_{cp}^{\mathrm{T}} \Phi_p^i z_p(N|k_j) \le h - \zeta_{c,N+i}^{[\tau]}, \quad i \in \mathbb{N}_{[0,\tau+\hat{N}]}, \\
& \sum_{p=1}^{N_p} g_{cp}^{\mathrm{T}} \Phi_p^i z_p(N|k_j) \le h - \bar{\zeta}_c^{[\tau]}, \quad i \in \mathbb{N}_{[\tau+\hat{N}+1,\tau+\hat{N}+n^*]},
\end{aligned}
\tag{4.35}
$$

The centralized self-triggered SMPC problem $\mathbf{P}_C^{\mathrm{ST}}(\mathbf{x}_c(k_j))$ at sampling time instant k_j is therefore formulated as

$$
\begin{aligned}
\tau_j^* &:= \max \left\{ \tau \in \mathbb{N}_{[1,\bar{\tau}]} \;\middle|\; \begin{array}{l} \mathcal{F}_C^{[\tau]}(\mathbf{x}_c(k_j)) \neq \emptyset, \\ V_C^{[\tau]}(\mathbf{x}_c(k_j)) \le V_C^{[1]}(\mathbf{x}_c(k_j)) \} \end{array} \right\}, \\
\mathbf{v}_C^*(k_j) &= \mathbf{v}_C^*(k_j;\tau_j^*),
\end{aligned}
\tag{4.36}
$$

where $\mathcal{F}_C^{[\tau]}(\mathbf{x}_c(k_j))$ is the feasible set to $\mathbf{P}_C^{[\tau]}(\mathbf{x}_c(k_j))$ and $V_C^{[\tau]}(\mathbf{x}_c(k_j))$ is the optimal value function to $\mathbf{P}_C^{[\tau]}(\mathbf{x}_c(k_j))$ defined as $V_C^{[\tau]}(\mathbf{x}_c(k_j)) = J_C(\mathbf{v}_C^*(k_j))$. The resulting centralized self-triggered SMPC method is summarized in Algorithm 3.

Theorem 6. *Under Algorithm 3, the self-triggered SMPC problem $\mathbf{P}_C^{\mathrm{ST}}(\mathbf{x}_c(k_j))$ is recursively feasible and the overall closed-loop system is quadratically stable as shown*

$$
\lim_{k_r \to \infty} \frac{1}{k_r} \sum_{k=0}^{k_r-1} \sum_{p=1}^{N_p} \mathbb{E}(\|x_p(k)\|_Q^2 + \|u_p(k)\|_R^2) \le \sum_{p}^{N_p} L_p.
\tag{4.37}
$$

Proof. (Recursive feasibility) At sampling time instant k_j, define τ_j^* and $\mathbf{v}_C^*(k_j)$ as the optimal solution to problem $\mathbf{P}_C^{\mathrm{ST}}(\mathbf{x}_c(k_j))$. The next sampling time instant is $k_{j+1} = k_j + \tau_j^*$. As discussed in Lemma 6, a feasible solution to $\mathbf{P}_C^{\mathrm{ST}}(\mathbf{x}_c(k_{j+1}))$

Algorithm 3: Centralized self-triggered SMPC algorithm.

Init.: For $p \in \mathcal{P}, c \in \mathcal{C}, \tau \in \mathbb{N}_{[1,\bar{\tau}]}$, determine local and coupled chance constraints tightening parameters $\beta_{p,i}^{[\tau]}$, $\zeta_{c,i}^{[\tau]}$ and terminal sets $\mathcal{Z}_{p,f}^{[\tau]}$ and $\mathcal{Z}_{c,f}^{[\tau]}$. Set $k = 0$.

while *Termination condition not satisfied* **do**

> **Step 1**: Measure the overall system state $\mathbf{x}_c(k)$;
>
> **Step 2**: Solve the self-triggered SMPC problem $\mathbf{P}_C^{\mathrm{ST}}(\mathbf{x}_c(k))$, obtain the optimal control input sequence $\mathbf{v}_C^*(k; \tau^*)$ for the overall system; broadcast the optimal inter-execution time interval τ^* and control input sequence $\mathbf{v}_p^*(k; \tau^*)$ to subsystem p;
>
> **Step 3**: Apply the control input $u_p(k+i) = K_p z_p(i|k) + v_p(i|k; \tau^*)$, for $i \in \mathbb{N}_{[1,\tau^*]}$;
>
> **Step 4**: Set the next sampling time instant $k = k + \tau^*$;
>
> **Step 5**: Go to Step 1.

end

---

includes

$$\tilde{\mathbf{v}}_C(k_{j+1}; 1) = [M_1^{\tau_j^*} \mathbf{v}_1^*(k_j; \tau_j^*) \dots M_p^{\tau_j^*} \mathbf{v}_p^*(k_j; \tau_j^*) \dots M_{N_p}^{\tau_j^*} \mathbf{v}_{N_p}^*(k_j; \tau_j^*)],$$

$$\tau_{j+1} = 1.$$

For $i \in \mathbb{N}_0$, it can be readily verified that

$$\sum_{p=1}^{N_p} g_{cp}^{\mathrm{T}} \Phi_p^i z_p(k_{j+1} + N|k_{j+1}) = \sum_{p=1}^{N_p} g_{cp}^{\mathrm{T}} \Phi_p^{\tau_j^* + i} z_p(k_j + N|k_j) + g_{cp}^{\mathrm{T}} \Phi_p^{N+i} \sum_{l=0}^{\tau_j^* - 1} A_p^l D_p w$$

$$\leq h - \zeta_{c,N+\tau_j^*+i}^{[\tau_j^*]} + b_{c,N+\tau_j^*+i}^{[\tau_j^*]} \leq h - \zeta_{c,N+i}^{[1]},$$

where the last inequality follows from Lemma 3.3 in [110]. Therefore, it can be concluded that a feasible solution exists for $\mathbf{P}_C^{\mathrm{ST}}(\mathbf{x}_c(k_{j+1}))$. By induction, $\mathbf{P}_C^{\mathrm{ST}}(\mathbf{x}_c(k_{j+l}))$ is feasible for $l \in \mathbb{N}_{\geq 1}$ if $\mathbf{P}_C^{\mathrm{ST}}(\mathbf{x}_c(k_j))$ is feasible.

(Stability) The stability proof follows the similar line of Theorem 4.2 in [110]. Define the cost function with $\tilde{\mathbf{v}}_C(k_{j+1}; 1)$ as $\tilde{V}_C^{[1]}(\mathbf{x}_c(k_{j+1}))$. Since the value of $\mathbf{x}_c(k_{j+1})$ is not known at time k_j, the expectation value of $\tilde{V}_C^{[1]}(\mathbf{x}_c(k_{j+1}))$ at k_j is given by $\mathbb{E}_{k_j}\{\tilde{V}_C^{[1]}(\mathbf{x}_c(\tau_j^*|k_j))\}$, where $\mathbf{x}_c(\tau_j^*|k_j)$ is the prediction of the augmented state $\mathbf{x}_c(k_{j+1})$ given $\mathbf{x}_c(k_j)$. By the design requirement $\alpha \geq 1$, it holds that $\frac{1}{\alpha} \sum_{p=1}^{N_p} \mathbb{E}_{k_j}\{\|x_p(\tau_j^*|k_j)\|_{Q_p}^2 + \|u_p(\tau_j^*|k_j)\|_{R_p}^2 - L_p\} \leq \sum_{p=1}^{N_p} \mathbb{E}_{k_j}\{\|x_p(\tau_j^*|k_j)\|_{Q_p}^2 + \|u_p(\tau_j^*|k_j)\|_{R_p}^2 - L_p\}$. Hence, from

(4.17), it holds that

$$\mathbb{E}_{k_j}\{\tilde{V}_C^{[1]}(\mathbf{x}_c(\tau_j^*|k_j))\} \leq V_C^{[\tau_j^*]}(\mathbf{x}_c(k_j)) - \sum_{p=1}^{N_p} \frac{1}{\alpha} \sum_{i=0}^{\tau_j^*-1} \mathbb{E}_{k_j}\{\|x_p(i|k_j)\|_{Q_p}^2 + \|u_p(i|k_j)\|_{R_p}^2 - L_p\}$$

$$\leq V_C^{[1]}(\mathbf{x}_c(k_j)) - \sum_{p=1}^{N_p} \frac{1}{\alpha} \sum_{i=0}^{\tau_j^*-1} \mathbb{E}_{k_j}\{\|x_p(i|k_j)\|_{Q_p}^2 + \|u_p(i|k_j)\|_{R_p}^2 - L_p\},$$

where the second inequality follows the triggering condition defined in (4.36). Furthermore, the optimality of $\mathbf{P}_C^{\mathrm{ST}}(\mathbf{x}_c(k_{j+1}))$ at time k_{j+1} implies that

$$\mathbb{E}_{k_j}\{V_C^{[1]}(\mathbf{x}_c(\tau_j^*|k_j))\} \leq V_C^{[1]}(\mathbf{x}_c(k_j)) - \sum_{p=1}^{N_p} \frac{1}{\alpha} \sum_{i=0}^{\tau_j^*-1} \mathbb{E}_{k_j}\{\|x_p(i|k_j)\|_{Q_p}^2 + \|u_p(i|k_j)\|_{R_p}^2 - L_p\}.$$

Summing the above inequality over $j \in \mathbb{N}_{[0,l]}$ and taking expectation on both sides leading to

$$\mathbb{E}_{k_l}\{V_C^{[1]}(\mathbf{x}_c(\tau_l^*|k_l))\} \leq \mathbb{E}_{k_0}\{V_C^{[1]}(\mathbf{x}_c(k_0))\}$$
$$-\frac{1}{\alpha} \sum_{p=1}^{N_p} \sum_{j=0}^{l} \sum_{i=0}^{\tau_j^*-1} \mathbb{E}_{k_j}\{\|x_p(i|k_j)\|_{Q_p}^2 + \|u_p(i|k_j)\|_{R_p}^2 - L_p\}.$$

Using the fact that $\mathbb{E}_{k_l}\{V_C^{[1]}(\mathbf{x}_c(\tau_l^*|k_l))\}$ is lower bounded leads to (4.37). Thus the proof is complete. $\square$

4.4 Distributed self-triggered stochastic MPC under coupled chance constraints

4.4.1 Distributed self-triggered SMPC algorithm

In the distributed self-triggered stochastic MPC setup, the centralized optimization problem $\mathbf{P}_C^{\mathrm{ST}}(\mathbf{x}_c(k))$ is distributed amongst subsystems as local optimization problems. In this work, the sequential update rule in [150] is adopted, meaning only one subsystem is required to update the control sequence at each time. At sampling time instant k_j, only subsystem p_k is permitted to solve the distributed optimization problem to get a new control sequence, while other subsystems $q \in \mathcal{Q}_{p_{k_j}}$ uses the constructed candidate control sequence:

$$\tilde{\mathbf{v}}_q(k_j; 1) := M_q^{\tau_{j-1}^*} \mathbf{v}_q^*(k_{j-1}; \tau_{j-1}^*)$$
$$= \left[v_q^*(\tau_{j-1}|k_{j-1}; \tau_{j-1}^*); \quad \cdots \quad ; \quad v_q^*(N-1|k_{j-1}; \tau_{j-1}^*); \mathbf{0} \right], \tag{4.38}$$

which is generated by augmenting the optimal control sequence at the previous sampling time instant k_j with $\mathbf{0}$. The update sequence $\{p_{k_0}, \ldots, p_{k_j}, \ldots\}$ is chosen by the designer and the cost function for subsystem $p = p_{k_j}$ with a fixed inter-execution time interval τ is defined as

$$J_p(\mathbf{v}_p(k_j; \tau)) = \frac{1}{\alpha} \sum_{i=0}^{\tau-1} \mathbb{E}(\|x_p(i|k_j)\|_{Q_p}^2 + \|u_p(i|k_j)\|_{R_p}^2 - L_p)$$
$$+ \sum_{i=\tau}^{\infty} \mathbb{E}(\|x_p(i|k_j)\|_{Q_p}^2 + \|u_p(i|k_j)\|_{R_p}^2 - L_p), \tag{4.39}$$

whose tractable expression can be found following Lemma 4. Hence, at sampling time instant k_j, the local optimization problem $\mathbf{P}_{D,p}^{[\tau]}(\mathbf{x}_c(k_j))$ for updating subsystem $p = p_{k_j}$ is defined as

$$
\begin{aligned}
\min_{\mathbf{v}_p(k_j; \tau)} \quad & J_p(\mathbf{v}_p(k_j; \tau)) \\
\text{s.t.} \quad & z_p(0|k_j) = x_p(k_j),\, z_q(0|k_j) = x_q(k_j), \\
& z_p(i+1|k_j) = \Phi_p z_p(i|k_j) + B_p v_p(i|k_j; \tau), \quad i \in \mathbb{N}_{[0,N-1]} \\
& g_p^{\mathrm{T}} z_p(i|k_j) \le h_p - \beta_{p,i}^{[\tau]}, \quad\quad\quad\quad\quad\quad\quad\; i \in \mathbb{N}_{[1,N-1]}, \\
& g_{cp} z_p(i|k_j) + \sum_{q \in \mathcal{P}_c} g_{cp} z_q^*(i|k_j) \le h_c - \eta_{c,i}^{[\tau]}, \quad i \in \mathbb{N}_{[1,N-1]} \\
& z_p(N|k_j) \in \hat{\mathcal{Z}}_{pc,f}^{[\tau]},
\end{aligned}
\tag{4.40}
$$

where the construction of the recursive feasible constraint tightening parameters $\beta_{p,i}^{[\tau]}$, $\eta_{c,i}^{[\tau]}$ and local terminal constraints $\hat{\mathcal{Z}}_{p,f}^{[\tau]}$ are given in subsection 4.3.2. $z_q^*(i|k_j), i \in \mathbb{N}_{[0,N]}$, denote the predicted nominal state for subsystem $q \in \mathcal{Q}_p$, which are determined by initial state $z_q(0|k_j)$ and candidate control sequence $\tilde{\mathbf{v}}_q(k_j; 1)$. It should be noted that values of $z_q^*(i|k_j)$ don not rely on $\mathbf{v}_p(k_j; \tau)$ and hence they are constant in the local optimization problem $\mathbf{P}_{D,p}^{[\tau]}(\mathbf{x}_c(k_j))$. The coupled terminal sets $\hat{\mathcal{Z}}_{pc,f}^{[\tau]}$ are defined as

$$
\hat{\mathcal{Z}}_{pc,f}^{[\tau]} := \left\{ z \;\middle|\; \begin{array}{ll}
g_p^{\mathrm{T}} \Phi_p^i z \leq h - \beta_{p,N+i}^{[\tau]}, & i \in \mathbb{N}_{[0,\tau+\hat{N}]}, \\[4pt]
g_p^{\mathrm{T}} \Phi_p^i z \leq h - \bar{\beta}_p^{[\tau]}, & i \in \mathbb{N}_{[\tau+\hat{N}+1,\tau+\hat{N}+n^*]}, \\[4pt]
g_{cp}^{\mathrm{T}} \Phi_p^i z \leq h - \zeta_{p,N+i}^{[\tau]} - \sum_{q \in \mathcal{P}_c} g_{cp}^{\mathrm{T}} \Phi_q^i z_q^*(N|k_j), & i \in \mathbb{N}_{[0,\tau+\hat{N}]}, \\[4pt]
g_{cp}^{\mathrm{T}} \Phi_p^i z \leq h - \bar{\zeta}_p^{[\tau]} - \sum_{q \in \mathcal{P}_c} g_{cp}^{\mathrm{T}} \Phi_q^i z_q^*(N|k_j), & i \in \mathbb{N}_{[\tau+\hat{N}+1,\tau+\hat{N}+n^*]}
\end{array} \right\},
$$

$$ \tag{4.41} $$

where $z \in \mathbb{R}^{n_{p,x}}$ and the parameter bound $\bar{\zeta}_p^{[\tau]}$ in (4.32b) is modified as

$$
\bar{\zeta}_c^{[\tau]} := \max_{i \in \mathbb{N}_{\geq \tau+1}, w \in \mathbb{W}} g_{cp} \Phi_p^{i-\tau} D_p \sum_{l=0}^{\tau-1} A_p^l D_p w + \sum_{q \in \mathcal{Q}_p} \max_{i \in \mathbb{N}_{\geq \tau+1}} g_{cq} \Phi_q^{i-\tau} D_q z_q^*(N|k_j)
$$

$$
+ \sum_{l=\tau+2}^{v_c-1} d_{c,l}^{[\tau]} + \sum_{p=1}^{N_p} \frac{\rho_p^{v_c}}{1-\rho_p} \|g_{cp}\|_{S_p} + \nu_{c,1}^{[\tau]}.
$$

Hence, the distributed self-triggered SMPC reformulation $\mathbf{P}_{D,p}^{\mathrm{ST}}(x_p(k_j))$ of the prototype self-triggered problem $\mathbf{P}_{O,p}^{\mathrm{ST}}(x(k_j))$ defined in (4.8) is given by

$$
\tau_j^* := \max \left\{ \tau \in \mathbb{N}_{[1,\bar{\tau}]} \;\middle|\; \begin{array}{l}
\mathcal{F}_{D,p}^{[\tau]}(x_p(k_j)) \neq \emptyset, \\[4pt]
V_p^{[\tau]}(x_p(k_j)) \leq V_p^{[1]}(x_p(k_j))
\end{array} \right\},
$$

$$
\mathbf{v}_p^*(k_j) = \mathbf{v}_p^*(k_j; \tau_j^*),
$$

$$ \tag{4.42} $$

where $\mathcal{F}_{D,p}^{[\tau]}(x_p(k_j))$ is the feasible set to $\mathbf{P}_{D,p}^{[\tau]}(x_p(k_j))$ and $V_p^{[\tau]}(x_p(k_j))$ is the optimal value function to $\mathbf{P}_{D,p}^{[\tau]}(x_p(k_j))$ defined as $V_p^{[\tau]}(x_p(k_j)) = J_p(\mathbf{v}_p^*(k_j; \tau_j^*))$. The resulting sequential distributed self-triggered SMPC algorithm is summarized in Algorithm 4.

4.4.2 Closed-loop properties of the distributed algorithm

The main result of this chapter is stated in the following theorem.

Theorem 7. *(Recursive feasibility and stability) At time instant $k_j = 0$, if centralized optimization problem $\mathbf{P}_C^{\mathrm{ST}}(\mathbf{x}_c(0))$ admits a feasible solution $\mathbf{v}_p^*(0)$ to subsystem $p \in \mathcal{P}$, and the closed-loop system is controlled under the Algorithm 4, then all subsequent distributed self-triggered optimization problems $\mathbf{P}_{D,p}^{\mathrm{ST}}(x_p(k_j))$ are feasible for $j > 0$. Furthermore, the closed-loop system satisfies the quadratic stability condition for the*

Algorithm 4: Cooperative distributed self-triggered SMPC algorithm

---

Init.: For $p \in \mathcal{P}, c \in \mathcal{C}, \tau \in \mathbb{N}_{[1,\bar{\tau}]}$, determine tightened local, global, and terminal constraints $\mathcal{Z}_{p,i}^{[\tau]}$, $\mathcal{Z}_{pc,i}^{[\tau]}$ and $\mathcal{Z}_{p,N}^{[\tau]}$. Determine the update sequence and set $k_0 = 0$. Solve the centralized optimization problem $\mathbf{P}_C^{\mathrm{ST}}(\mathbf{x}_c(k_0))$, obtain control sequence $\mathbf{v}_p^*(k_0; \tau_0^*)$ and inter-execution time τ_0^*. For each subsystem p, apply $u_p(k_0 + i) = K_p z_p(k_0 + i) + v_p^*(i|k_0; \tau_0^*)$, $i \in \mathbb{N}_{[0,\tau_0^*]}$, set sampling time instant as $k = k_0 + \tau_0^*$, and construct candidate input $\tilde{\mathbf{v}}_q(k) = M_q^{\tau_0^*} \mathbf{v}_q^*(k_0; \tau_0^*)$.

while *Termination condition not satisfied* **do**

    Step 1: For update system p_k:

       1. Recieve measurement $x_q(k)$ from $q \in \mathcal{Q}_{p_k}$;

       2. Obtain $\mathbf{v}_{p_k}^*(k)$ and τ_k^* by solving $\mathbf{P}_{D,p}^{\mathrm{ST}}(x_{p_k}(k))$;

       3. Apply $u_{p_k}(k + i) = K_{p_k} z_p(k + i) + v_{p_k}^*(i|k)$ for $i \in \mathbb{N}_{[0,\tau_k^*]}$ and transmit the inter-execution time τ_k^* to other subsystem $q \in \mathcal{Q}_{p_k}$;

    Step 2: For subsystem $q \in \mathcal{Q}_{p_k}$:

       1. Transmit $x_q(k)$ to subsystems p_k and recieve τ_k^*;

       2. Apply $u_q(k + i) = K_q z_q(k + i) + \tilde{v}_q(i|k)$ for $i \in \mathbb{N}_{[0,\tau_k^*]}$

       3. Construct $\mathbf{v}_q(k + \tau_k^*) = M_q^{\tau_k^*} \mathbf{v}_q(k)$;

    Step 3: Set $k = k + \tau_k^*$, and return to Step 1.

end

---

entire system as shown by

$$\lim_{k_r \to \infty} \frac{1}{k_r} \sum_{k=k_0}^{k_r-1} \sum_p^{N_p} \mathbb{E}\{\|x_p(k)\|_Q^2 + \|u_p(k)\|_R^2\} \leq \sum_{p=1}^{N_p} L_p. \qquad (4.43)$$

Proof. (Recursive feasibility) If the centralized self-triggered MPC $\mathbf{P}_C^{\mathrm{ST}}(\mathbf{x}_c(0))$ is initially feasible, then there exit feasible solutions $\{\mathbf{v}_p^*(k_0; \tau_0^*)\}$ for subsystem $p \in \mathcal{P}$. At the next sampling time instant $k_1 = \tau_0^*$, denote the update system as p_{k_1} and define $\tilde{\mathbf{v}}_{p_{k_1}}(k_1) = M_{p_{k_1}}^{\tau_0^*} \mathbf{v}_{p_{k_1}}^{\tau_0^*}(k_0)$. It can be readily verified $\tilde{\mathbf{v}}_{p_{k_1}}(k_1)$ satisfies the local chance constraint (4.29a). The coupled chance constraint (4.24b) can be ensured if the centralized problem $\mathbf{P}_C^{[1]}(\mathbf{x}_c(k_1))$ adopts $z_q^*(k_1) = x_q(k_1), z_{p_{k_1}}(k_1) = x_{p_{k_1}}(k_1)$, and $\tilde{\mathbf{v}}_q^1(k_1) = M_q^{\tau_0^*} \mathbf{v}_q^{\tau_0^*}(k_0)$. Therefore, it follows that $\tilde{\mathbf{v}}_{p_{k_1}}(k_1)$ is a feasible solution to the distributed optimization problem $\mathbf{P}^{[1]}_{D,p_{k_1}}(x_{p_{k_1}}(k_1))$ at time instant k_1, which implies that the problem $\mathbf{P}^{\mathrm{ST}}_{D,p}(x_{p_{k_1}}(k_1))$ is feasible.

Next, let the feasible solution at sampling time instant k_{j-1} be $\{\mathbf{v}_p(k_{j-1}; \tau^*_{k_{j-1}})\}$ for all subsystems $p \in \mathcal{P}$. Following the Algorithm 4, the candidate control input sequence will be updated to $\tilde{\mathbf{v}}_q(k_j) = M_q^{\tau^*_{j-1}} \mathbf{v}_q(k_{j-1}; \tau^*_{j-1})$ for all subsystems $q \neq p_{k_j}$. For subsystem p_{k_j}, the distributed self-triggered MPC problem $\mathbf{P}^{[1]}_{D,p_{k_j}}(x_{p_{k_j}}(k_j))$ is equivalent to the centralized self-triggered MPC problem $\mathbf{P}^{[1]}_{C}(\mathbf{x}(k_j))$ constrained to $\mathbf{v}_q(k_j) = \tilde{\mathbf{v}}_q(k_j)$. From the recursive feasibility of $\mathbf{P}^{\mathrm{ST}}_{C}(\mathbf{x}(k_j))$ in Theorem 6, it can be verified that $\tilde{\mathbf{v}}_q(k_j), p \in \mathcal{P}$ are feasible solutions to $\mathbf{P}^{\mathrm{ST}}_{C}(\mathbf{x}(k_j))$ and $\tilde{\mathbf{v}}_{p_{k_j}}(k_j) = M_{p_{k_j}}^{\tau^*_{j-1}} \mathbf{v}_{p_{k_j}}(k_{j-1}; \tau^*_{j-1})$ is feasible to $\mathbf{P}^{\mathrm{ST}}_{D,p_{k_j}}(x_{p_{k_j}}(k_j))$. Therefore, it implies that initial feasible solutions to $\mathbf{P}^{\mathrm{ST}}_{C}(\mathbf{x}(k_0))$ ensures $\mathbf{P}^{\mathrm{ST}}_{D,p_{k_j}}(x_{p_{k_j}}(k_j)), j \in \mathbb{N}_0$ are feasible regardless of update sequence.

(Quadratic stability proof) At sampling time instant $k_j, j \in \mathbb{N}_0$, define the update subsystem as p_{k_j}, and denote the solution to corresponding distributed self-triggered optimization problem $\mathbf{P}^{\mathrm{ST}}_{D,p_{k_j}}(x_{p_{k_j}}(k_j))$ as $\tau^*_{k_j}$ and $\mathbf{v}^*_{p_{k_j}}(k_j; \tau^*_{k_j})$. Let $J_{p_{k_j}}(\mathbf{v}_{p_{k_j}}(k_j; \tau^*_{k_j}))$ be a stochastic Lyapunov function candidate at sampling time instant k_j for update system p_{k_j}. The global cost over the whole system is defined as the summation of cost associated with each subsystem $J_C(\mathbf{v}_C(k_j; \tau^*_{k_j})) = \sum\limits_{p=1}^{N_p} J_p(\mathbf{v}_p(k_j; \tau^*_{k_j}))$. At sampling time instant k_{j+1}, a feasible candidate solution for subsystem p_{k_j} is defined as $\tilde{\mathbf{v}}_{p_{k_j}}(k_{j+1}; 1) = M_{p_{k_j}}^{\tau^*_{k_j}} \mathbf{v}_{p_{k_j}}(k_j; \tau^*_{k_j})$ with $\tau_{k_{j+1}} = 1$.

At sampling time instant k_{j+1}, the updating subsystem is $p_{k_{j+1}}$ and the closed-loop system follows Algorithm 4. Thus, the global cost at time k_{j+1} is

$$J_C(\mathbf{v}_C(k_{j+1}; \tau^*_{k_{j+1}})) = J_{p_{k_{j+1}}}(\mathbf{v}_{p_{k_{j+1}}}(k_{j+1}; \tau^*_{k_{j+1}})) + \sum_{q \neq p_{k_{j+1}}}^{N_p} J_q(M_q^{\tau^*_{k_{j+1}}} \mathbf{v}_q(k_j; \tau^*_{k_j})).$$

$$(4.44)$$

Define $\tilde{J}_C(\tilde{\mathbf{v}}_C(k_{j+1}; \tau^*_{k_{j+1}}))$ as the cost with all subsystem updating with the candidate control $\tilde{\mathbf{v}}_p(k_{j+1}; 1) = M_p^{\tau^*_{k_j}} \mathbf{v}_p(k_j; \tau^*_{k_j})$, $p \in \mathcal{P}$. It follows that

$$\begin{aligned}
J_C(\mathbf{v}_C(k_{j+1}; \tau^*_{k_{j+1}})) =& \tilde{J}_C(\tilde{\mathbf{v}}_C(k_{j+1}; \tau^*_{k_{j+1}})) \\
&+ J_{p_{k_{j+1}}}(\mathbf{v}_{p_{k_{j+1}}}(k_{j+1}; \tau^*_{k_{j+1}})) - J_{p_{k_{j+1}}}(\tilde{\mathbf{v}}_{p_{k_{j+1}}}(k_{j+1}; 1)).
\end{aligned}$$

$$(4.45)$$

From the optimality of $\mathbf{P}^{\mathrm{ST}}_{D,p_{k_{j+1}}}(x_{p_{k_{j+1}}}(k_{j+1}))$, it follows that

$$J_{p_{k_{j+1}}}(\mathbf{v}_{p_{k_{j+1}}}(k_{j+1};\tau^*_{k_{j+1}})) - J_{p_{k_{j+1}}}(\tilde{\mathbf{v}}_{p_{k_{j+1}}}(k_{j+1};1)) \leq 0,$$

which further implies that $J_C(\mathbf{v}_C(k_{j+1};\tau^*_{k_{j+1}})) \leq \tilde{J}_C(\tilde{\mathbf{v}}_C(k_{j+1};\tau^*_{k_{j+1}}))$. From the triggering condition in (4.42), it holds $J_{p_{k_{j+1}}}(\mathbf{v}_{p_{k_{j+1}}}(k_{j+1};\tau^*_{k_{j+1}})) \leq J_{p_{k_{j+1}}}(\mathbf{v}_{p_{k_{j+1}}}(k_{j+1};1))$. So we have

$$\begin{aligned}
\mathbb{E}\{J_C(\mathbf{v}_C(k_{j+1};1))\} &\leq J_C(\mathbf{v}_C(k_j;1)) \\
&\quad -\frac{1}{\alpha}\sum_{i=0}^{\tau^*_j-1}\sum_p^{N_p}\mathbb{E}\{\|x_p(i|k_j)\|_Q^2 + \|u_p(i|k_j)\|_R^2 - L_p\}.
\end{aligned} \qquad (4.46)$$

Summing (4.46) for $0 \leq j \leq r$ and taking expectation on both sides lead to

$$\begin{aligned}
\mathbb{E}\{J_C(\mathbf{v}_C(k_r;1))\} &\leq J_C(\mathbf{v}_C(k_0;1)) \\
&\quad -\sum_{j=0}^{r-1}\frac{1}{\alpha}\sum_{i=0}^{\tau^*_j-1}\sum_p^{N_p}\mathbb{E}\{\|x_p(i|k_j)\|_Q^2 + \|u_p(i|k_j)\|_R^2 - L_p\}.
\end{aligned} \qquad (4.47)$$

Since $J_C(\mathbf{v}_C(k_0;1))$ is finite and $\alpha \geq 1$ by assumption and $\mathbb{E}\{J_C(\mathbf{v}_C(k_r;1))\}$ is lower bounded, it holds that

$$\lim_{r\to\infty}\frac{1}{k_r}\sum_{j=0}^{r-1}\sum_{i=0}^{\tau^*_j-1}\sum_p^{N_p}\mathbb{E}\{\|x_p(i|k_j)\|_Q^2 + \|u_p(i|k_j)\|_R^2\} \leq \sum_{p=1}^{N_p}L_p, \qquad (4.48)$$

which implies the quadratic stability condition that

$$\lim_{k_r\to\infty}\frac{1}{k_r}\sum_{k=k_0}^{k_r-1}\sum_p^{N_p}\mathbb{E}\{\|x_p(k)\|_Q^2 + \|u_p(k)\|_R^2\} \leq \sum_{p=1}^{N_p}L_p. \qquad (4.49)$$

Thus the proof is complete. $\qquad\square$

4.5 Numerical examples

4.5.1 Case 1: Homogeneous subsystems

In the first example, we consider the control of a group of distributed homogeneous DC-DC converter systems subject to additive disturbance and coupled chance constraint. The benchmark DC-DC converter model has been widely utilized as in [48], [110]. Model parameters in (4.1) are given as for $\forall p \in \mathbb{N}_{[1,3]}$:

$$A_p = \begin{bmatrix} 1 & 0.0075 \\ -0.143 & 0.996 \end{bmatrix}, B_p = \begin{bmatrix} 4.798 \\ 0.115 \end{bmatrix}, D_p = \begin{bmatrix} 1 & 0 \\ 0 & 1 \end{bmatrix}.$$

The additive disturbance $w_p(k)$ for subsystem p is assumed to be independently and identically truncated Gaussian distributed with zero mean and variance 0.04^2, and the bound in (4.2) on $w_p(k)$ is $\alpha_{p,i} = 0.1$, $i = 1, 2$. Local and coupled chance constraints on system states are given as:

$$g_p = \begin{bmatrix} 1 & 0 \end{bmatrix}, h_p = 2, p_p = 0.8,$$
$$g_{cp} = \begin{bmatrix} 1 & 0 \end{bmatrix}, h_c = 5.5, p_{p,c} = 0.8, \forall p \in \mathbb{N}_{[1,3]}, c = 1.$$

The prediction horizon is defined as $N = 8$. To construct the terminal constraint (4.33) and (4.34), the extended prediction horizon is chosen as $\hat{N} = 12$ and $n^* = 0$. The tuning parameter and weighting matrices in the cost function (4.39) are selected as $\alpha = 1.2$ and $Q_p = \begin{bmatrix} 1 & 0; 0 & 3 \end{bmatrix}$, $R_p = 1$. The linear feedback gain K_p in (4.4) is chosen as $K_p = \begin{bmatrix} 0.263 - 0.329 \end{bmatrix}$, which is LQ optimal, and the maximal inter-execution interval is chosen as $\bar{\tau} = 3$. The simulation is conducted in Matlab 2019b with Yalmip [153] and the QP solver is Gurobi [154]. The initial condition for each subsystem is defined as $x_1(0) = \begin{bmatrix} 3.5 & 3 \end{bmatrix}^T, x_2(0) = \begin{bmatrix} 2.5 & 2 \end{bmatrix}^T$ and $x_3(0) = \begin{bmatrix} 2.5 & 2.8 \end{bmatrix}^T$. The simulation length is $T_{sim} = 18$ steps. As discussed in Section 4.4, the sequential update sequence $\{1, 2, 3, \dots\}$ is adopted to update the control action for each subsystem. For comparison purposes, simulations with 1000 realizations of disturbances are performed for the proposed distributed self-triggered SMPC (DSTSMPC), distributed self-triggered robust MPC (DSTRMPC, setting $p_p = 1, p_{p,c} = 1$) and distributed SMPC (DSMPC [110]).

Chance constraints satisfaction: Figure 4.3 demonstrates the closed-loop trajectories $\{x_p(k), k = 1, \dots, T_{sim}\}$ for each subsystem and sum of states $\{\sum_p^{N_p} x_p(k), k = $

$1, \ldots, T_{sim}\}$ of system (4.1) controlled by Algorithm 4. To demonstrate constraint violations, regions around the constraint bound are enlarged as shown in the right two figures in Figure 4.3. From simulation results, at time $k = 1$, 16%, 0%, 19% of the closed-loop trajectories of subsystem $1, 2, 3$ violate the local constraints (4.3a) and 14%, 0%, 18% at time step $k = 2$, 17%, 0%, 19% at time step $k = 3$. The coupled constraint (4.3b) violation probability is 12% at $k = 1$, 11% at $k = 2$, and 15% at $k = 3$. The simulation results demonstrate that the convergence of the system state to a region around the origin and the constraint violation probability satisfies the specific requirement.

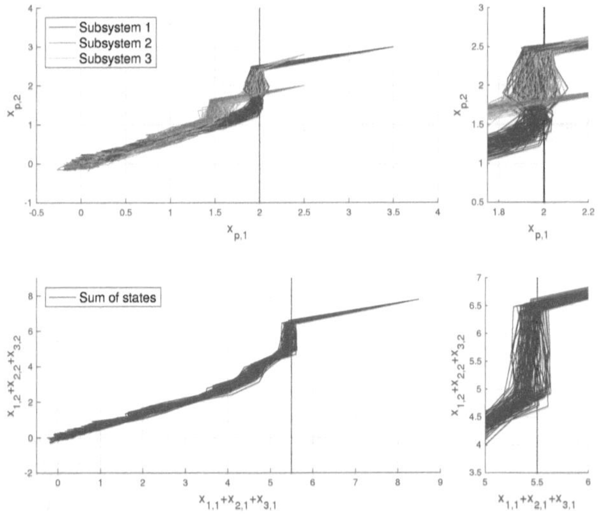

Figure 4.3: Top figures show closed-loop trajectories of homogeneous subsystems under the sequential update rules. The vertical black line is the state constraint for three subsystems. The blue, green and red lines denote the state trajectories of subsystem 1,2,3, respectively. Bottom figures show the evolution of sum of all states. Right figures show the enlarged region around the constraint bound.

Average communication reduction and performance evaluation: Figure 4.4 demonstrates the sum of closed-loop trajectories of overall systems with one realization of disturbance under DSTSMPC, DSTRMPC and DSMPC. Red markers in the line denote sampling steps, and the amount of sampling instants has been reduced significantly. In addition, it can be observed that no constraint violation occurs for the DSTRMPC scheme. The average communication time between each subsystem of

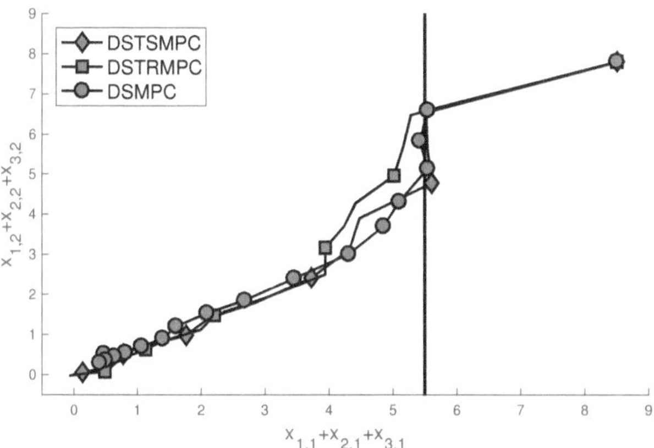

Figure 4.4: Evolution of $\sum_{p=1}^{3} x_p(k)$ under DSTSMPC, DSTRMPC and DSMPC for 1 realization of the uncertainty sequence.

the distributed self-triggered SMPC is $\tau_{aver} = 2.8$, which implies an average communication reduction by 59.9% compared to the distributed SMPC scheme. Moreover, define the performance index as

$$J_{index} = \frac{1}{T_{sim}} \sum_{p=1}^{N_p} \sum_{k=0}^{T_{sim}} (\|x_p(k)\|_{Q_p}^2 + \|u_p(k)\|_{R_p}^2 - L_p).$$

It can be shown that J_{index} is 16.71 for DSTSMPC, 16.31 for DSMPC, and 18.75 for DSTRMPC. It concludes that the communication between each subsystem is reduced significantly without sacrificing too much performance. This can also be observed from Figure 4.4 since all closed-loop trajectories converge to the origin in a similar pattern. By allowing for constraint violation, the performance of the proposed DSTSMPC is improved compared to the DSTRMPC scheme.

4.5.2 Case 2: Heterogeneous subsystems

Consider a team of heterogeneous subsystems, which has been used in [90] and [155], modeled by

$$
\begin{aligned}
A_1 &= \begin{bmatrix} 1.6 & 1.1 \\ -0.7 & 1.2 \end{bmatrix}, & B_1 &= \begin{bmatrix} 1 \\ 1 \end{bmatrix}, \\
A_2 &= \begin{bmatrix} 1.5 & 1.1 \\ 0 & 1.2 \end{bmatrix}, & B_2 &= \begin{bmatrix} 0.8 \\ 0.9 \end{bmatrix}, \\
A_3 &= \begin{bmatrix} 1.4 & 1.2 \\ -0.3 & 1.1 \end{bmatrix}, & B_3 &= \begin{bmatrix} 1.2 \\ 0.8 \end{bmatrix}, \\
D_p &= \begin{bmatrix} 1 & 0 \\ 0 & 1 \end{bmatrix}, & p &\in \mathcal{P} := \mathbb{N}_{[1,3]},
\end{aligned}
$$

and the chance constraints are defined as

$$
\begin{aligned}
g_1 &= \begin{bmatrix} 1 & 1.3 \end{bmatrix}, & h1 &= 15, & p_1 &= 0.8, \\
g_2 &= \begin{bmatrix} 1.4 & 0.6 \end{bmatrix}, & h2 &= 8.4, & p_2 &= 0.8, \\
g_3 &= \begin{bmatrix} 0.9 & 0.4 \end{bmatrix}, & h3 &= 9, & p_3 &= 0.8, \\
g_{cp} &= g_p, & h_c &= 33, & p_{p,c} &= 0.8, p \in \mathcal{P}.
\end{aligned}
$$

The disturbance $w_{p,i}(k), i = 1, 2$ follows a truncated Gaussian distribution with zero mean and variance $\dfrac{1}{12^2}$ and $|w_{p,i}(k)| \leq 0.5$. The subsystem control update sequence is defined as $\{1, 2, 3, \ldots, \}$ and $N = 6$, $\hat{N} = 7$, $n^* = 1$, $Q_p = I_{2\times 2}$, $R = 1$, $\alpha = 1.3$. The feedback gain K_p are chosen as the LQ optimal gain as $K_1 = \begin{bmatrix} -1.04 & -1.04 \end{bmatrix}$, $K_2 = \begin{bmatrix} -0.93 & -1.19 \end{bmatrix}$, $K_3 = \begin{bmatrix} -0.76 & -0.95 \end{bmatrix}$. The initial state for each subsystem is given by $x_1(0) = \begin{bmatrix} -6 & 25 \end{bmatrix}^T$, $x_2(0) = \begin{bmatrix} -3 & 45 \end{bmatrix}^T$, $x_3(0) = \begin{bmatrix} -6 & 60 \end{bmatrix}^T$, and 1000 simulations are carried out with different realizations of uncertainties w_p.

The closed-loop trajectories of each subsystem is illustrated in Figure 4.5. At time step $k = 1$, 19%, 20%, 18% of the closed-loop trajectories violate the local constraints (4.3a) for subsystem $1, 2, 3$, respectively. When simulation step length is selected as $T_{sim} = 10$ and the self-triggered tuning parameter is chosen as $\alpha = 2$, the average communication time during the transient is $\tau_{aver} = 1.8$, implying that over 30% communication between each subsystem are reduced. The average performance index J_{index} for DSTSMPC is 1438.65 while it is 1390.90 for DSMPC.

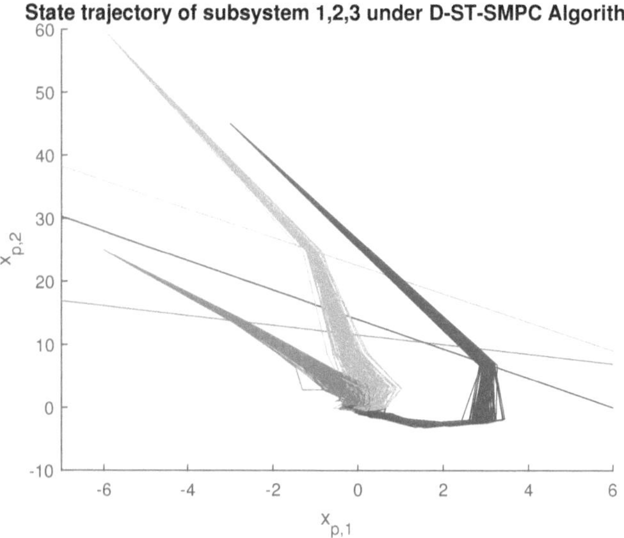

Figure 4.5: The red, blue and greed lines denote the closed-loop state trajectories of subsystem 1,2,3, respectively.

4.6   Conclusions

In this chapter, a distributed self-triggered stochastic MPC control scheme is proposed for CPSs subject to chance constraints and additive disturbances. To be more specific, the self-triggered SMPC proposed in [110] is extended to distributed paradigm subject to coupled chance constraints. The communication burden between each subsystem can be significantly reduced while guaranteeing chance constraints satisfaction. Both local and coupled chance constraints are transformed into the deterministic form using the constraints tightening method in [151]. In addition, sufficient conditions to guarantee recursive feasibility of the algorithm and stability of the closed-loop system are developed. The results are illustrated by numerical examples for homogeneous and heterogeneous systems.

Chapter 5

Conclusions

The book mainly focuses on the control analysis and synthesis of aperiodically sampled stochastic model predictive control schemes. We design appropriate SMPC strategies for different types of stochastic systems subject to various uncertainties and disturbances, and rigorously analyze the resulting closed-loop properties. We further demonstrate the effectiveness of the proposed SMPC methods through com-prehensive numerical examples.

5.1 Conclusions

In Chapter 2, a stochastic self-triggered MPC scheme is proposed for linear constrained discrete-time systems. The proposed self-triggered sampling scheme effectively reduces the communication load between the sensor and the controller thanks to the implementation of the self-triggered sampling scheme. The recursive feasibility of the proposed control scheme and the stability conditions are developed. Simulation results have demonstrated the effectiveness of the algorithm.

In Chapter 3, a novel self-triggered SMPC algorithm with adaptive prediction horizon is proposed for linear systems subject to additive disturbances and state chance constraints. The prediction horizon in the MPC algorithm changes adaptively to generate appropriate inter-execution time intervals. To deal with the additive disturbance, an improved triggering condition is designed and the asymptotic sampling behavior is analyzed. Sufficient conditions to guarantee the recursive feasibility of the algorithm are given, and the closed-loop system is proven to be quadratical stable. Simulation results have shown the efficacy of designed self-triggered control method in reducing the communication burden while guaranteeing some specific performance loss.

In Chapter 4, a distributed self-triggered stochastic MPC control scheme is proposed for CPSs subject to coupled probabilistic constraints and additive disturbances. To be more specific, the self-triggered SMPC proposed in [110] for a single system is extended to distributed systems subject to coupled constraints. The communication burden among subsystems can be significantly reduced while guaranteeing probabilistic constraints satisfaction. Both local and coupled probabilistic constraints are transformed into the deterministic form using the constraints tightening method in [151]. In addition, sufficient conditions to guarantee recursive feasibility of the algorithm and stability of the closed-loop system are developed.